DON'T DIE YOUNG

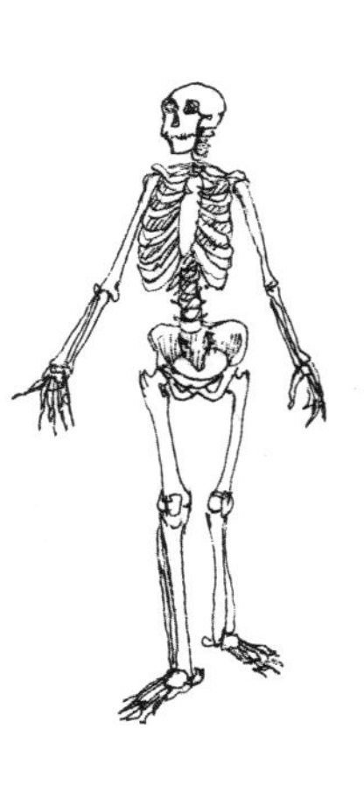

DON'T DIE YOUNG

an anatomist's guide to your organs and your health

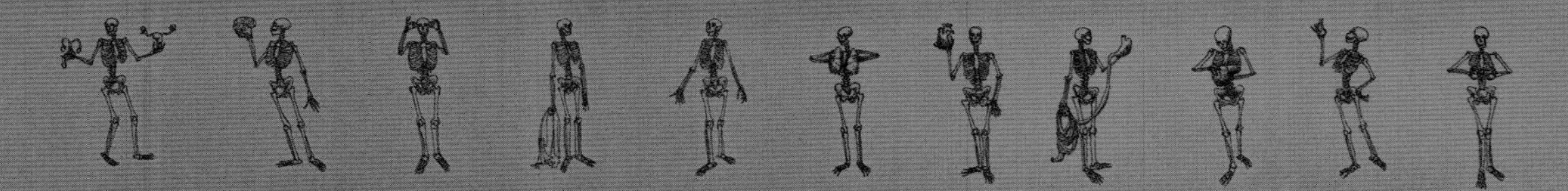

Dr Alice Roberts

BLOOMSBURY

The great leap forward [in Medicine] is a comprehensive and unprecedented understanding of what the healthy body is and how it survives and protects itself.

Jonathan Miller, *The Body in Question*

First published in Great Britain in 2007

Skeleton illustrations by Alice Roberts
For picture credits see page 271

Editor: Ann Grand
Designer: Smith & Gilmour, London
Picture researcher: Juliet Davis
Illustrator: Oxford Designers and Illustrators Ltd
Indexer: David Atkinson

Bloomsbury Publishing Plc, 36 Soho Square, London W1D 3QY

A CIP catalogue record for this book is available from the British Library.

ISBN: 9780747590255

10 9 8 7 6 5 4 3 2 1

Printed in Great Britain by Butler & Tanner Limited, Frome

All papers used by Bloomsbury Publishing are natural, recyclable products made from wood grown in well-managed forests. The manufacturing processes conform to the environmental regulations of the country of origin.

www.bloomsbury.com

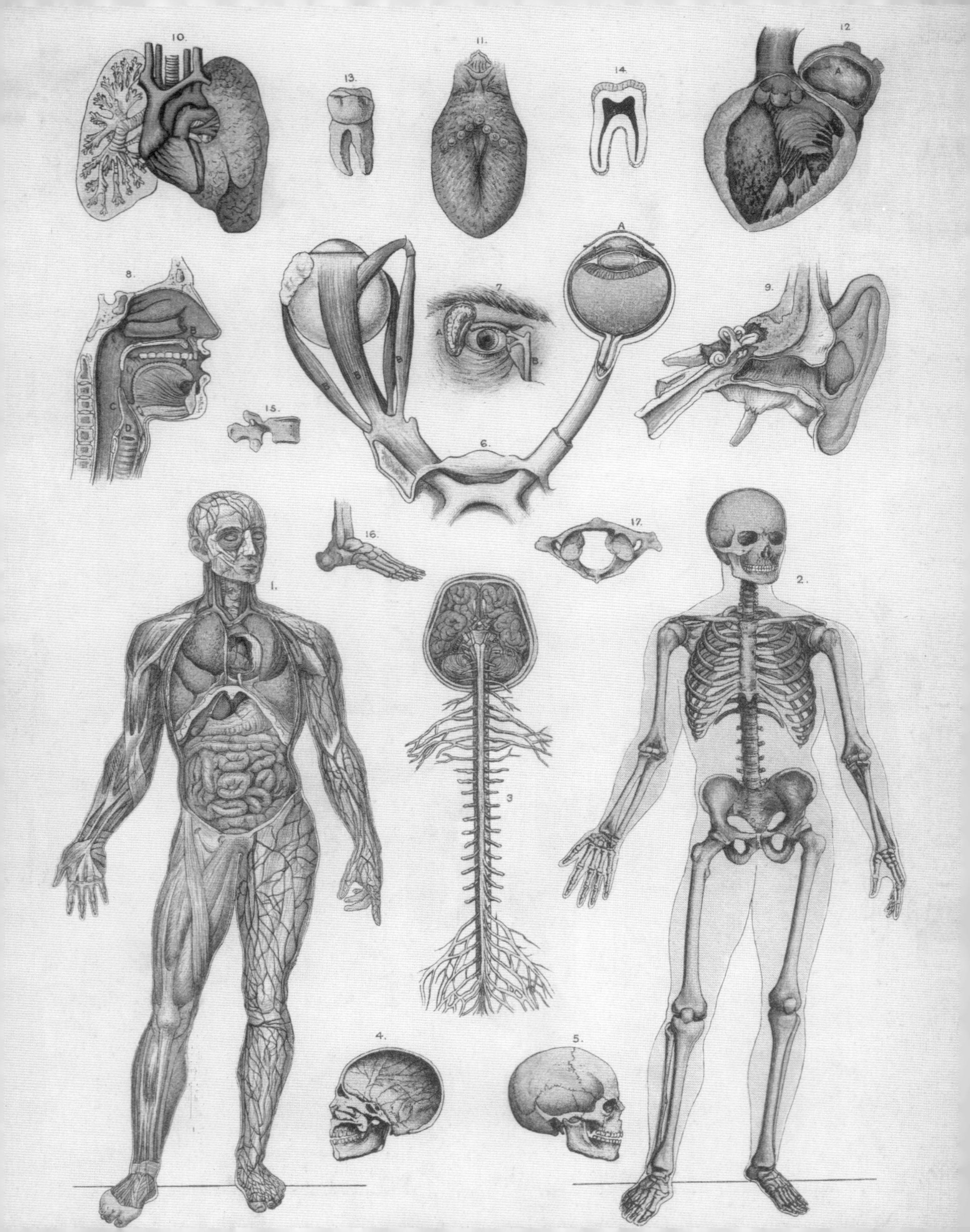
10.
11.
13.
14.
12.
A
A
8.
B
7
A
B.
9.
B
B
B
C
D
15.
6.
16.
17.
1.
2.
3
4.
5.

Foreword

Can you cheat death? No one's managed it yet, although a French woman called Jeanne Calment clocked up 122 years and 164 days, which is pretty good going. She was a vegetarian, and stuffing yourself with fruit and vegetables (via the mouth) is currently one of the best ways of ensuring a longer life. But she must also have had lucky genes.

I occasionally see patients who defy the odds, notching up a century while smoking 40 woodbines a day and living on a diet of sugar, and their DNA must have an amazing capacity to repair itself in the face of such a toxic onslaught. Most people who play health roulette aren't so lucky. As Dr Alice Roberts observes in this excellent book: 'Eight out of ten cases of heart disease, nine out of ten cases of acquired diabetes and three in ten cases of cancer could be avoided by diet and lifestyle changes'.

But instead of just telling you what to do to avoid an early death, Dr Roberts tells you why. Most people haven't experienced the joy and wonder of dissecting a fellow human, but this is the next best thing. We're taken on an anatomical roller coaster through all the major organ systems, with clear explanations of what can go wrong and how to put it right. It's written with enthusiasm and humour, and a good mix of anecdote and research.

The final chapter pulls it all together and identifies common themes. The inevitable truth is that we're all rusting to death, some of us more quickly than others. We think of the oxygen we breathe as life giving, but it is actually very toxic. It reacts hungrily with proteins and enzymes, stopping them from working. It burns us in a Faustian pact; oxygen is essential to our life, but it also destroys living tissue and eventually kills us.

We may not be able to live for ever, but the trick is to avoid an early death and to keep the human machine running smoothly enough to enjoy our life. We only go around the track once, but if everyone digested this book there'd be far fewer breakdowns on the way.

Dr Phil Hammond

A range of body parts from a nineteenth-century anatomical illustration.

Introduction

The human body is a marvellous machine. Within this machine, labour is divided among large components, each specialized to carry out a particular job. There are components that operate like bellows, to draw oxygen into the body, and there's another that acts as a pump to move oxygen and nutrients around the body. There's a fuel converter that takes big chunks of all sorts of plant and animal matter and breaks them down into small compounds that can be burned for energy or recycled as building materials. There's a factory which takes in raw materials and either uses them immediately as building blocks, or stores them for future use. The human machine has a waste management system, where rubbish is cleaned up and stored, ready for disposal. It even has components that are designed to make new generations of human machines.

This machine is aware of its environment: it has special parts that respond to photons and allow it to build up a moving picture of the outside world, parts that monitor compression waves in the air and parts that can detect liquids or minute amounts of chemicals in the air. It is protected from the environment by a membrane that covers its entire outside, a membrane that can sense the temperature and textures that are around it.

The human machine can move itself around. And it can think: a central processing unit makes it fully interactive. Working together, all these components, the organs of the human body, strive to keep us alive.

The organs are packed neatly inside the body, like pieces in a three-dimensional jigsaw, nestled into crevices and interlocked with each other. Also packed into the body's inner spaces are the service ducts and communication system of the body: the arteries, veins, lymphatic vessels and nerves. Around and between these organs, vessels and nerves are layers of packaging: connective tissue that keeps the parts of the body in place and protects them. The organs are separate entities – you can surgically remove an eye, the liver, the heart, a lung – and each has its

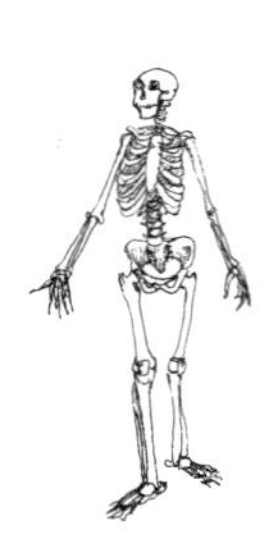

own specialist job. But together, the whole becomes more than the sum of its parts: the individual organs depend on each other. The body is healthy when all its organs are working at their best.

Health means different things to different people. To some, it implies a feeling of well-being, a sense that all the parts of the body are working to their best ability. Doctors tend to view health differently, as an absence of illness. I will try to engage with both of these philosophies: health as a positive experience of the body working well, and health as the absence of disease. The first view depends on an understanding of the structure of the organs and how they function, when they're functioning well. There are ways that you can live your life that will nurture your organs, by treating them well and making sure they're getting what they need to work properly. The second view starts with the problems that arise when organs malfunction and regards healthy living as a way of avoiding the circumstances which contribute to these problems. In each chapter, I shall look at the healthy organ and how it functions. Then I'll turn my attention to a few of the common problems that may affect the organ, in particular to conditions that can be influenced by making changes in lifestyle and diet.

Although many, many studies have shown the positive effects of a healthy diet and lifestyle, it's clear that not many of us manage to live our lives in a way that will give us the best chances of health, long life and happiness. This may be partly to do with the barrage of advice we get about the minutiae of how we live and what we eat: what type of exercise to do and how often, how many apricot kernels to eat each week or which molecule-with-an-unbelievably-long-name we should endeavour to include in our diets. Newspapers pick up on studies about how an obscure foodstuff might be helpful to some aspect of health, and trumpet each as the *Next Big Thing*, the elixir of life. What I'd like to do is to explain some of the stuff you read about in newspapers or hear on the news. (What exactly are antioxidants and omega-3 fish oils? Are they really good for you? And if so, where can you find them without having to buy pills?) There is an ocean of information and misinformation out there, which can sometimes make it difficult to find out which life and dietary factors are the most important for keeping healthy. I've waded through part of the morass to find advice for good health that is based on good science and firm evidence.

Boiled down, a lot of the advice is really common sense. Keeping healthy is actually a lot easier than some people would have you believe. If you're not already sceptical, you should be; especially if someone tries to sell you a 'health' product but can't tell you how it works or that it's not possible to decide if it works by scientific methods. Health is BIG business, so lots of people have a large stake in getting you to believe their notion of how to achieve perfect health. (We're very lucky, in Britain, to have a healthcare system that doesn't depend on selling itself to you. If you're not paying your GP every time you see them, he or she doesn't have any financial incentive to keep you coming back.)

This book takes a no-nonsense approach to health. Rest assured; I don't have any pills or energy bars to sell you, I'm not a faith healer and I won't stick tubes up your bottom. What I would like to do is help you understand how your body works and how to look after it, hopefully at no added expense.

Organ by organ, we'll look at the structure and function of the body and how to keep it as healthy as possible. I'll end each chapter with a few important recommendations for lifestyle and diet relating to that particular organ, based on official guidelines and the latest research. Living a healthy lifestyle gives your body the best possible chances for avoiding illness and staying in good health. It doesn't mean that you won't ever get ill, because there are too many other factors involved in determining that – not least the genes you are born with – but your choice of diet and lifestyle will certainly affect how long, happy and healthy your life is.

Page 12 **Code for a human: chromosomes seen with the aid of an electron microscope, magnified over 10,000 times.**

Page 13 **Spindly nerve fibres: these particular neurons bring sensory messages from the body to the spinal cord. They have been stained with a dye that shows up the proteins which form a scaffold inside each cell and give it its shape.**

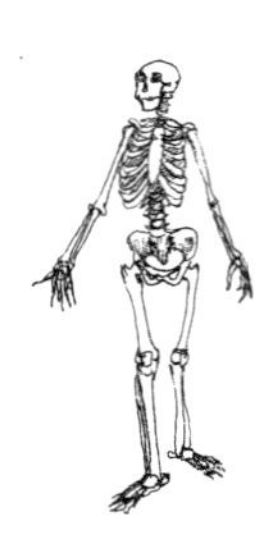

The building blocks of the body

At the most basic level, the human body is a collection of atoms. It's not a random collection; the atoms (carbon, hydrogen, oxygen and many other elements) make up molecules such as sugars, proteins, oils and, of course, the molecule that holds the code for the body itself – deoxyribonucleic acid, DNA. The molecules are then arranged to form cells, each with its own copy of the DNA code.

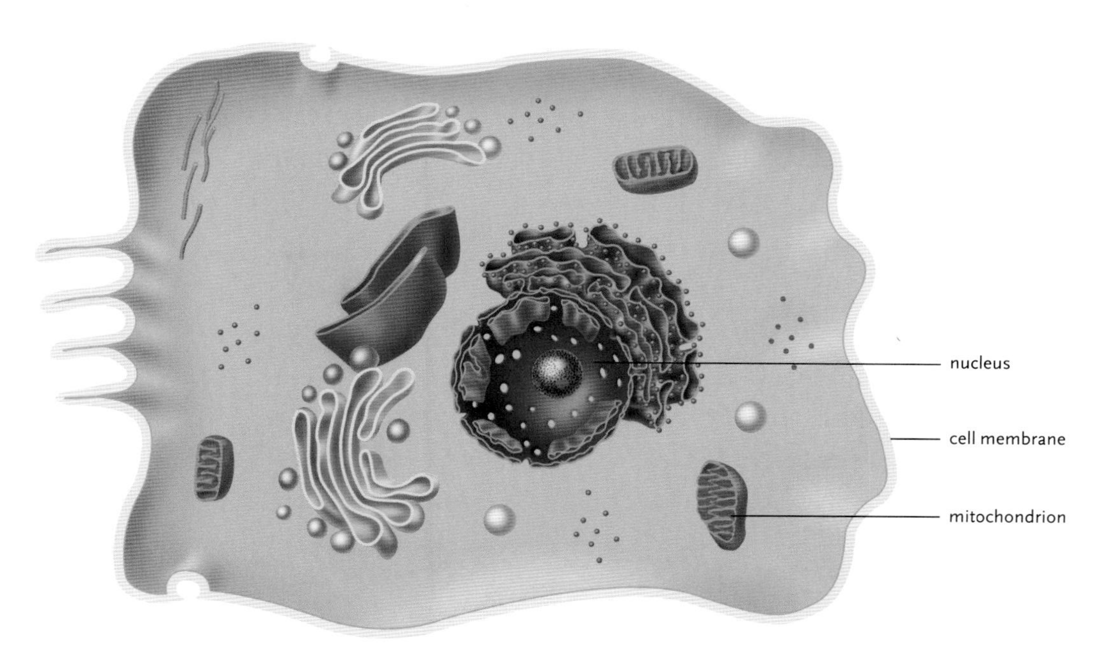

Specialized packages inside a cell.

If you could zoom in and look closely at a cell, you would see that specialization is not just something that happens at the large-scale level of the organism and its organs: every tiny cell contains specialized packages doing different jobs. One package, the nucleus, houses the database that the cell uses to run its daily functions.

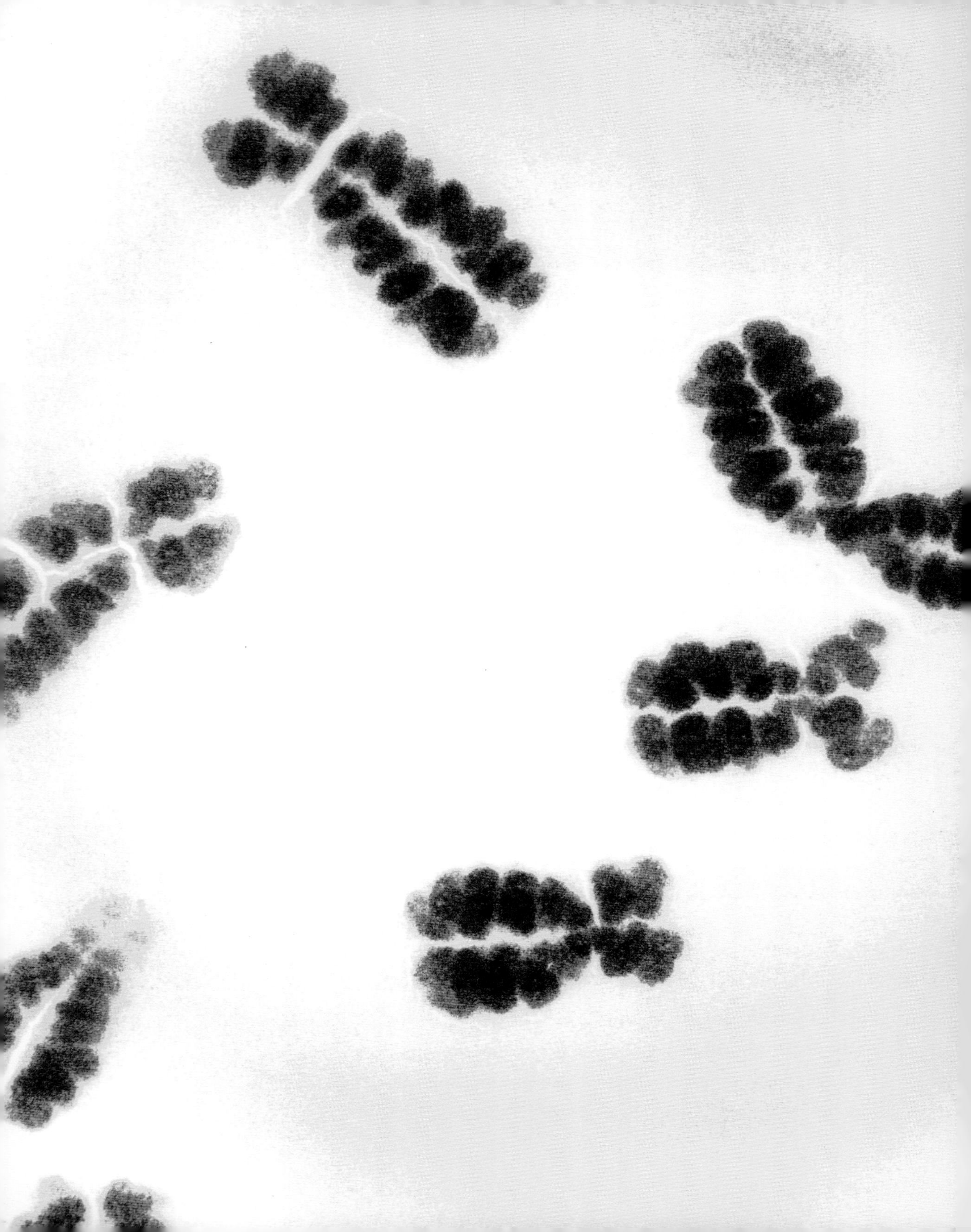

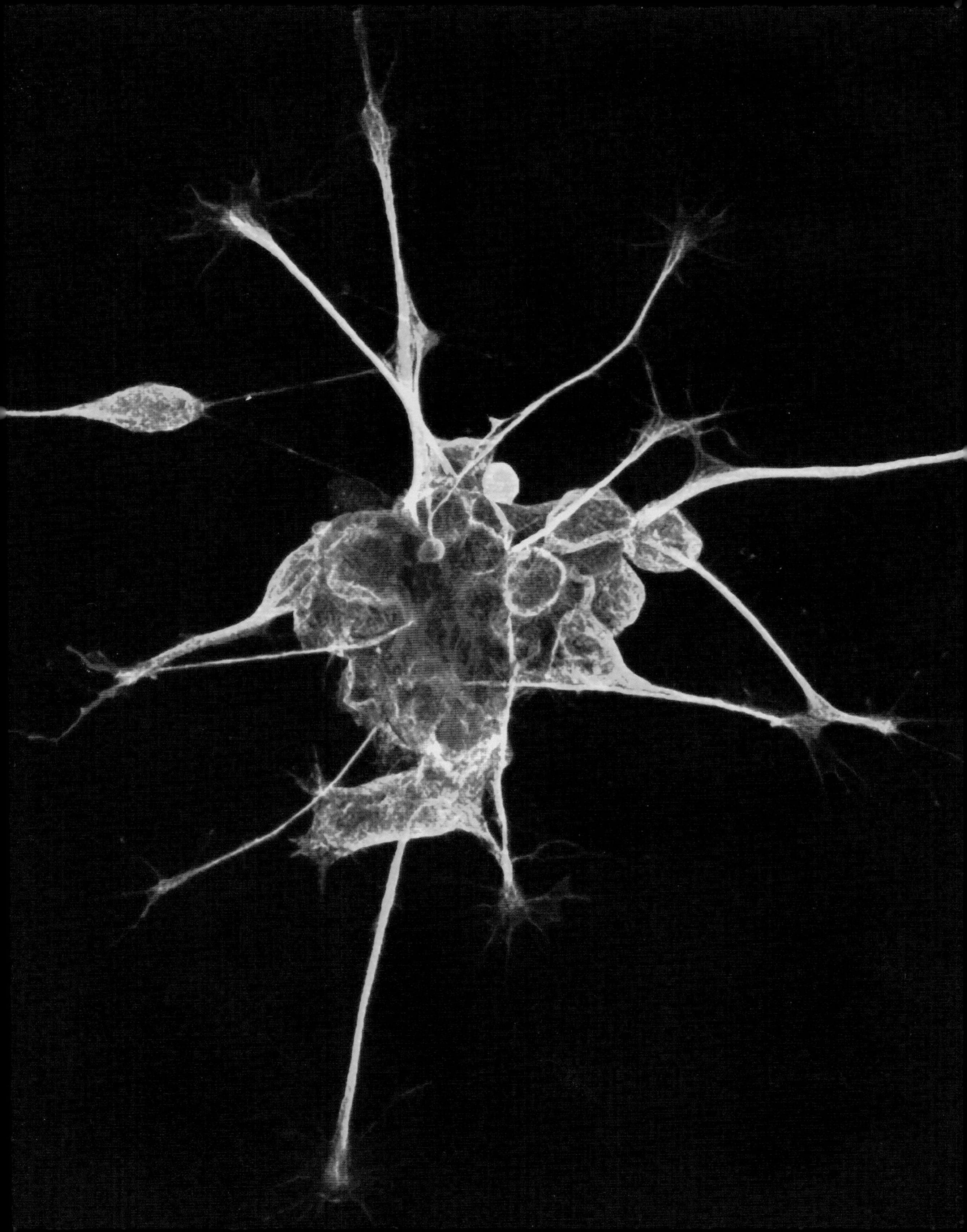

The database is written as long strings of DNA, in 23 pairs of chromosomes. The DNA strings are, essentially, a library of coded instructions on how to make all the proteins the cell might ever need, as well as lots it never will. Every cell in the body has the same library but by switching on different sets of genes, many different cell types can be made, from the spindly nerve fibre to the plump fat cell, the octopus-like podocyte in the kidney or the rod-like cells in the retina of the eye: they each have the same library but they're choosing to read different books. There are also packages where messages from the DNA inside the nucleus are 'translated' into new proteins; as if the DNA were sending out orders to factories to make specific products. Other packages contain destructive enzymes that the cell uses to protect itself.

The mitochondria are the power stations of the cell, taking fuel – sugar – and burning it to produce energy. This energy is used to charge a molecular battery: adenosine triphosphate (ATP). This unglamorous-sounding chemical is absolutely essential to life. Wherever and whenever the cell needs energy to perform a task, it uses ATP.

These packages are basic characteristics of human cells, or cells in any animal, but beyond the basics, cells are very diverse in look and function, depending on which books they read from their DNA library. Nerve cells (neurons) have a long thin projection called the axon, which is like a minute electric cable along which the nerve impulses travel. Muscle cells (myocytes) are either small, spindle-shaped cells (found in smooth or involuntary muscle – the type inside organs and the wall of the intestine) or long rods, made of lots of cells fused together (in striated muscle, the stuff that makes up your biceps and your quads). Red blood cells are very unusual; they don't have a nucleus (in fact they lose their nucleus as they develop) and are really just bags of haemoglobin, doing its job of transporting oxygen around the body. Bone cells (osteocytes) and cartilage cells (chondrocytes) are little, round cells. The retina of the eye has very weirdly shaped cells, the rods and cones, which respond to light falling on them. Most cells stay pretty much stationary, apart from the red blood cells and white blood cells.

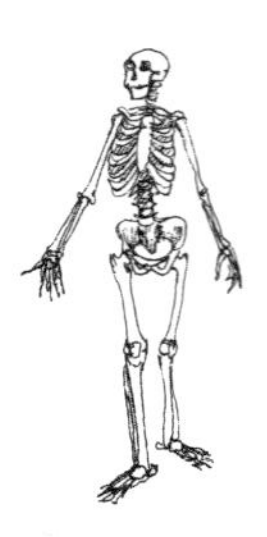

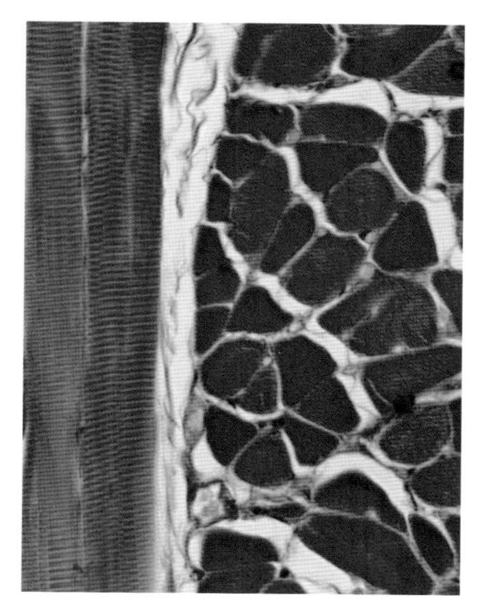

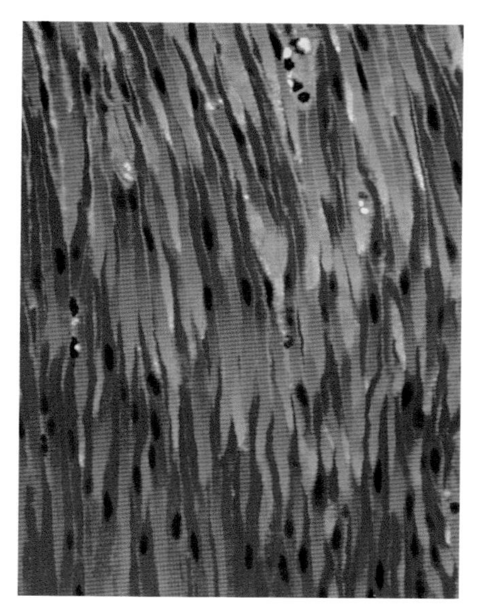

Left **Fibres of voluntary or striated muscle, like biceps or quadriceps (the striations are clear in the two fibres on the left, cut along their length).**

Right **Spindle-shaped cells in involuntary or smooth muscle, like the type in the gut wall. Each dark spot is the nucleus of one myocyte.**

Tissues

Tissues are collections of cells. A collection of neurons forms nervous tissue. A collection of myocytes forms muscle tissue. Many tissues contain something else, called a 'matrix', as well as cells, which might be a fluid, a gel or a harder substance. These 'connective tissues' are a bit like a currant bun or blueberry muffin: the fruity bits are the cells and the cakey bit is the matrix. Tissues with a matrix include blood (where blood cells float in a fluid matrix), cartilage (where chondrocytes rest in a gel matrix) and bone (where osteocytes sit in tiny caves inside a hard, mineralized matrix).

A more generalized connective tissue contains fibroblast (Greek for 'fibre-sprouting') cells in a matrix made of collagen and elastin, which are both proteins that form long fibres. If the fibres are lined up in parallel to each other, the result is a tissue that is good at resisting tension – this is what tendons and ligaments are made of. If the collagen

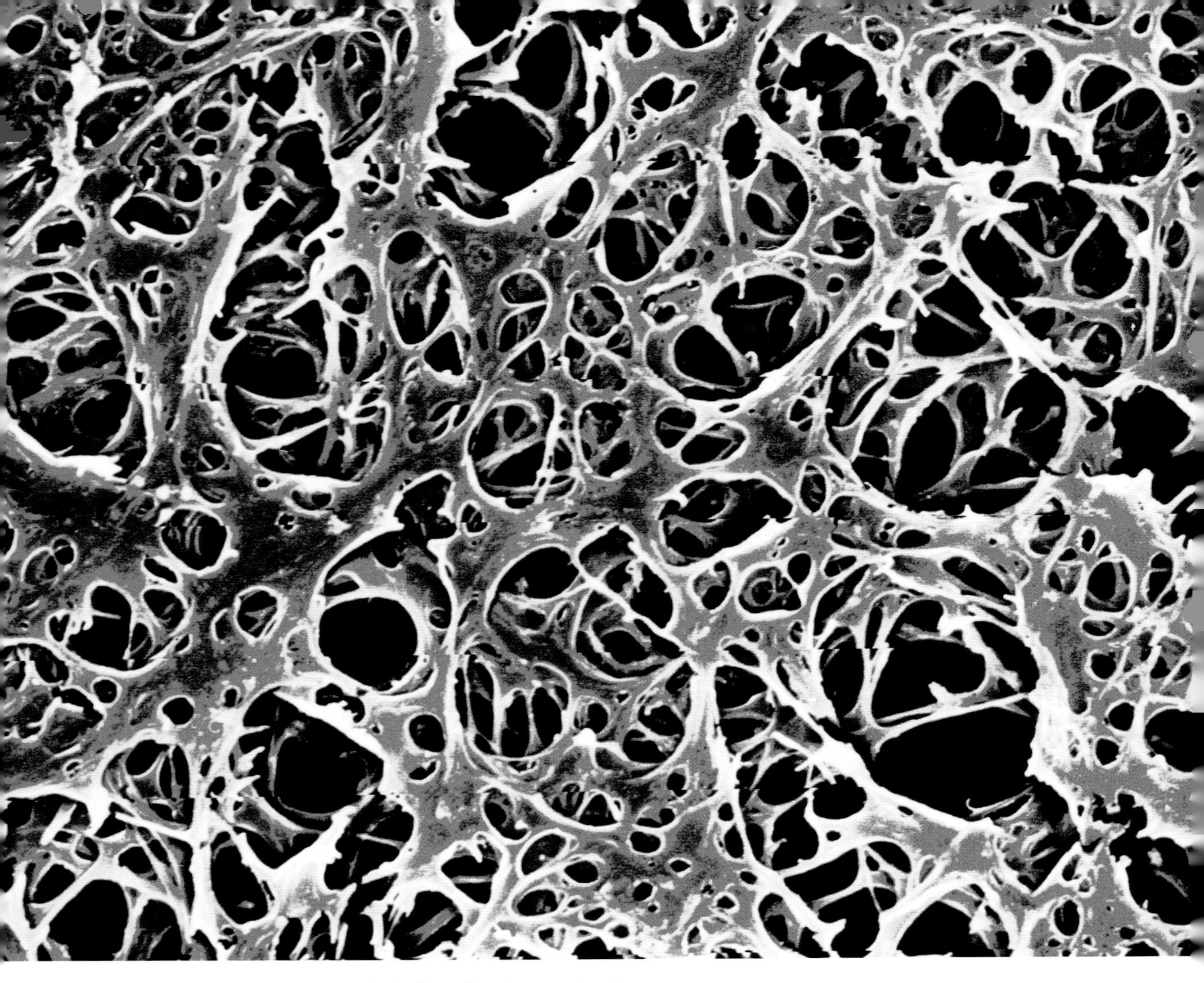

A network of collagen fibres in connective tissue.

fibres are arranged in a haphazard way, the tissue is 'loose connective tissue'. This can be very varied: in some places, it is quite dense and fibrous, in other areas it has a cobweb-like wispiness, and in yet others, it might be packed with fat cells. This is the stuff that anatomists refer to as 'fascia', which means 'swaddling' or 'bundling' – it's the packaging of the body. When surgeons operate, they often follow 'fascial planes': rather than tearing through the layers of fascia, they will open up its natural planes in the tissue with their fingers or a blunt instrument, peeling apart the layers. In the dissection room, fascia gets in the way

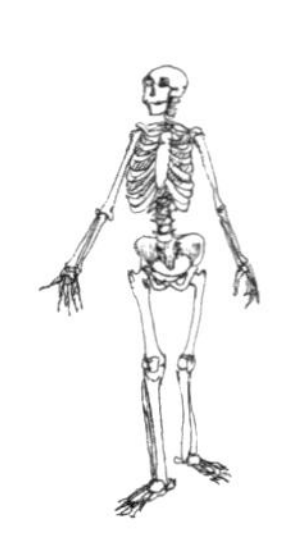

of seeing organs and vessels, so anatomists will carefully clean fascia away to reveal the contents inside the packaging.

Some tissues form barriers or distinct linings on the inside or the outside of the body. These 'epithelia' usually consist of a sheet of cells (which might be as little as one cell thick or several storeys high) sitting on a membrane, called the basement membrane. A special type of epithelium, the endothelium, lines the inside of the blood vessels. The smallest blood vessels, the capillaries, are tiny tubes of endothelium just about wide enough to squeeze one red blood cell through.

The endothelial wall is very, very thin, which means substances can easily pass through the walls of the capillaries. So, in your lungs, oxygen passes from the air into the red blood cells in the capillaries; in the tissues, oxygen passes back out of the capillaries; in the gut, dissolved nutrients pass into the capillaries.

Organs

Organs are collections of different tissues. Every organ has blood vessels: from large ones, with smooth muscle walls, lined with endothelium on the inside and connective tissue on the outside, down to tiny endothelial capillaries. They also have bundles of nerve fibres, wrapped up in connective tissue sheaths. Some nerves carry messages away from the organ to the brain, with information about pain or stretching (and about the external environment in the case of skin). Other nerves carry messages to the organ, instructing it to stimulate cells to release secretions or to make smooth muscle in the organ contract. There may be ducts, lined with epithelium, inside the organ. There will also be specialized tissues made of cells that are very specific to the organ: liver cells in the liver, heart muscle cells in the heart, light-sensitive cells in the eye, and cells that divide to form eggs in the ovary or sperm in the testes.

Looking inside the body

The study of the structure of the body is anatomy, which literally means 'to cut up'. Since ancient times, surgeons have learnt about the body and its organs by dissection. Young surgeons often take jobs as anatomy demonstrators, because they know that teaching anatomy to medical students will help them brush up their own knowledge, ready for surgical exams. Preparing dissections for teaching allows them to get a really good three-dimensional appreciation of the body and its contents, to find out exactly where organs, vessels and muscles are.

I've always thought of this way of learning as similar to the way mechanics learn their trade. It's all very well looking at books but the mechanic needs to strip an engine to get a real feel for how it all fits

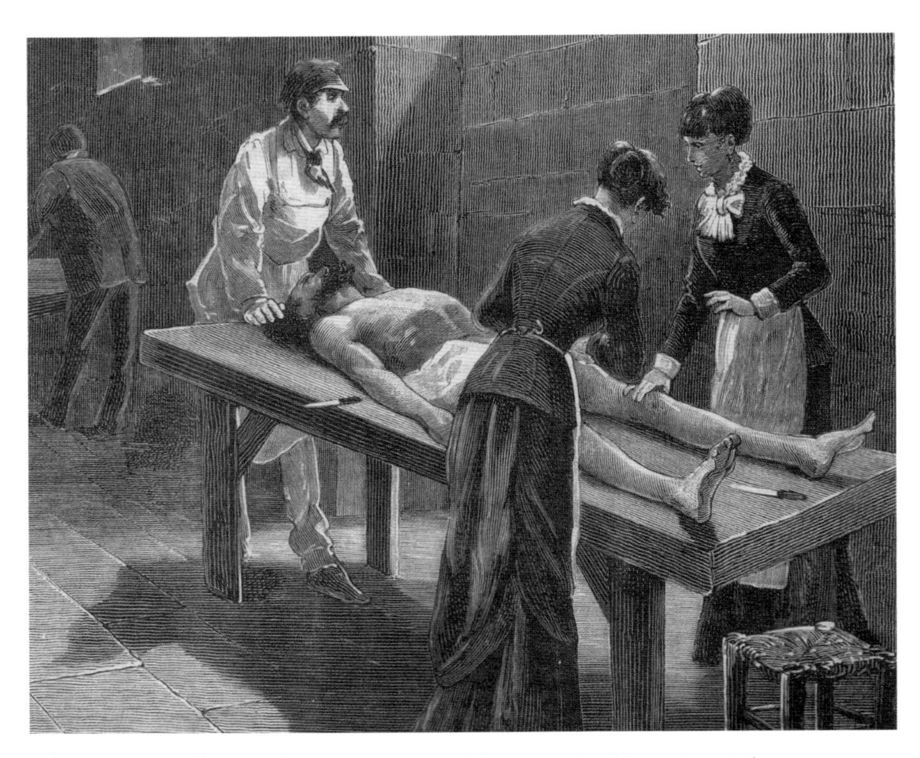

A time-honoured study: human dissection was resurrected as a way of studying anatomy in the Renaissance. This nineteenth-century engraving shows students about to embark on their study of anatomy by dissecting a cadaver. Cadaveric dissection still underpins the study of anatomy today.

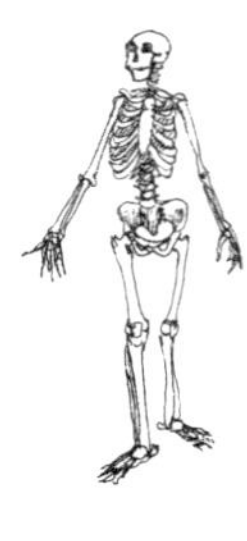

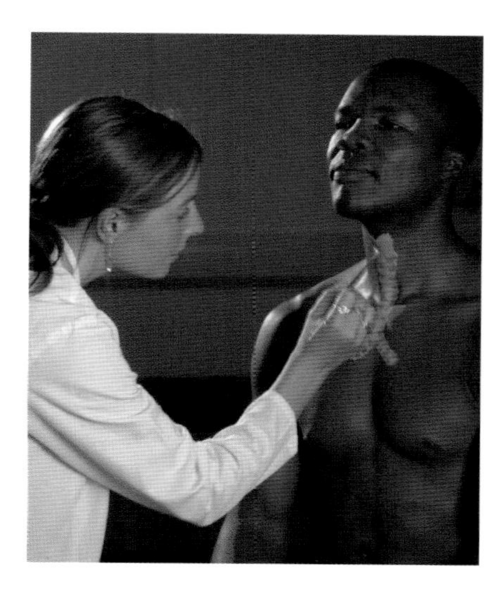
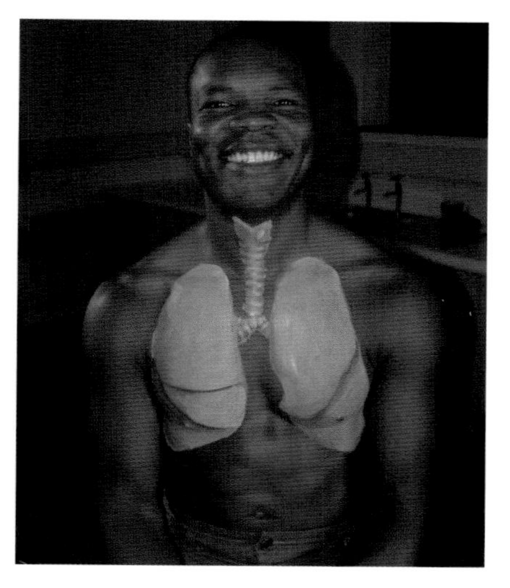

Being able to imagine the organs on the surface, as though they had 'X-ray vision', allows doctors to examine their patients. I sometimes get medical students to draw on each other to help visualize the anatomy inside. Here, I'm illustrating the surface anatomy of the lungs by painting the lungs on the climber, Trevor Massiah.

together – there's something about doing it yourself and using your hands that you don't get from books or computer animations. There are some great plastic and computer-based anatomical models available but they don't show the amazing intricacy of the human body, the variation between people or give you the feel of the lowly packing tissue. Human dissection still has a place in medical learning in the twenty-first century, and medical schools are very grateful to those generous people who donate their bodies for this purpose. It's an amazing gift to humanity: the bequest of your own body to teach the next generation of doctors.

I've been privileged to learn human anatomy through dissecting the body and seeing operations, as an undergraduate medical student, a practising doctor, a medical demonstrator and an anatomy lecturer. I've been able to explore the complexity of the structure of the human body, and its intricate and beautiful design. (I use the word 'design' advisedly, as the human body has not been designed by a third party: it's a product of evolution, and it got to be the way it is over millions of years of tweaking.) The human body is a wonderful thing, and I hope that this book will convey some of that wonder.

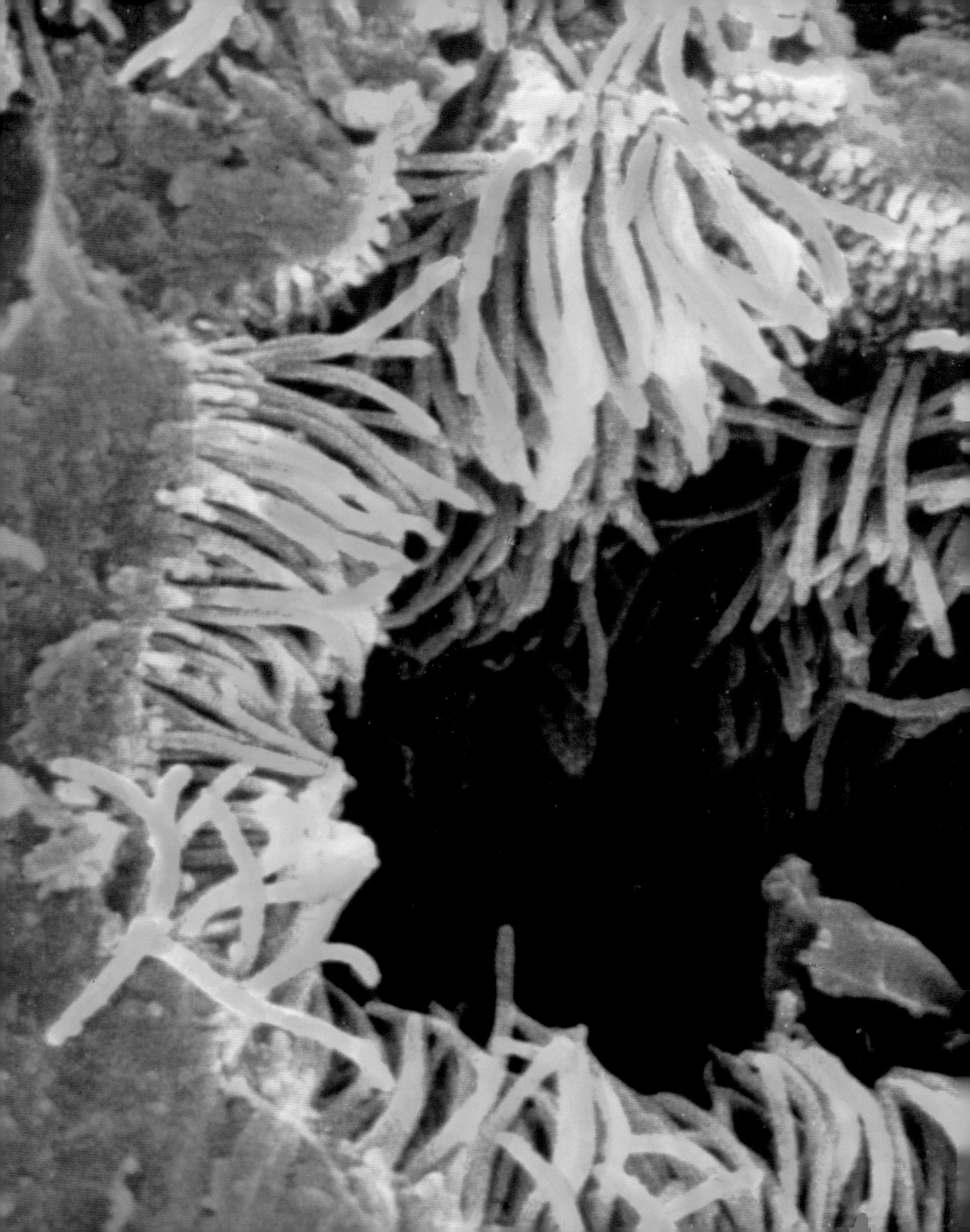

ONE

THE LUNGS

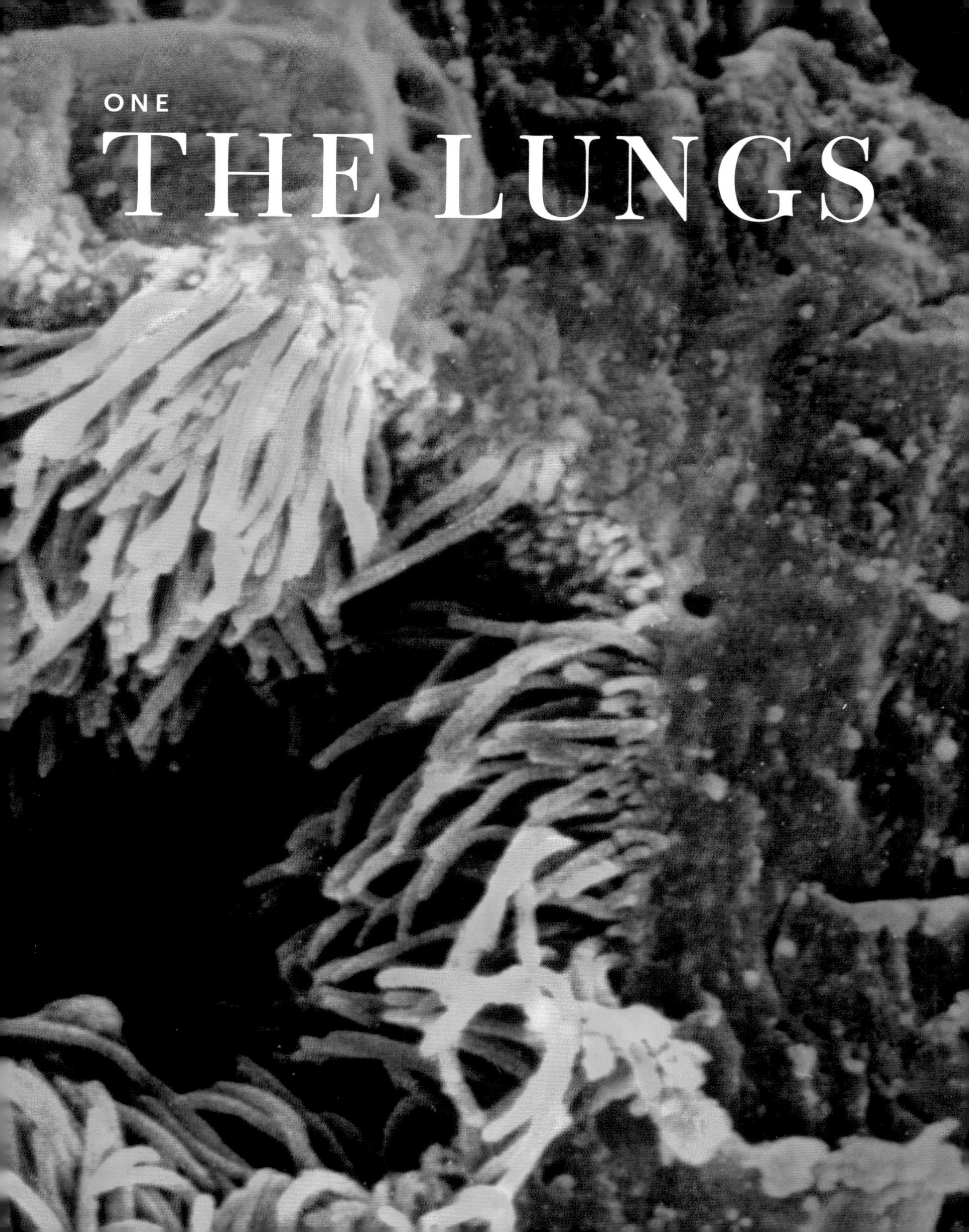

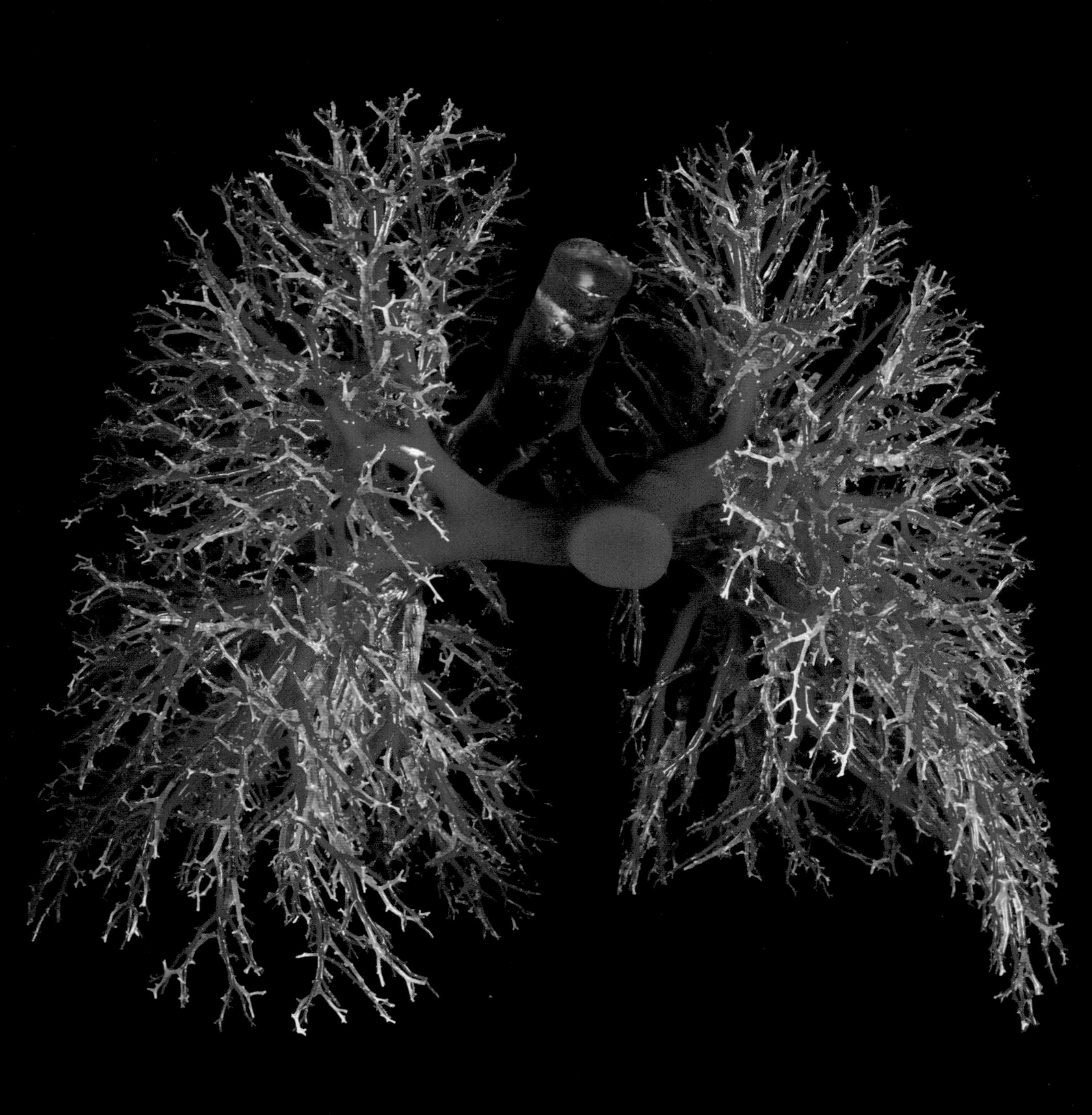

Every cell in our bodies needs to get oxygen in – and carbon dioxide out. In tiny animals, like the single-celled amoeba, these gases simply diffuse through the cell membrane, into the cell. Larger animals need to transport these gases around the body in the blood. The transfer of gases between air and blood happens in the lungs. 'Lung' is a good old Anglo-Saxon word, and shares a root with the word 'light': lungs are light because they are full of air. Butchers refer to animals' lungs as their 'lights'.

The lungs are essentially a pair of elastic bags that can expand to pull air inside the body. Every time we breathe in, air is drawn into the lungs and oxygen passes into millions of tiny capillaries – ready to be carried off to the rest of the body. Simultaneously, carbon dioxide diffuses out of the blood and into the lungs' air spaces, ready to be breathed out.

The air spaces of the lungs are tiny and numerous: each lung contains about 300 million tiny air bubbles, the 'alveoli', where the exchange of oxygen and carbon dioxide happens. If you could open up and spread out your alveoli, they would cover a huge surface area of up to 80 square metres.

Previous pages **Self-cleaning epithelium: under the electron microscope, a carpet of tiny hairs or cilia (green) is seen on surface of cells lining the airways. These cilia waft mucus and trapped dirt out of the lungs.**

Left **In this resin cast, the airways in the lungs have been filled with clear resin, and the branches of the pulmonary artery in red; the rest of the lung tissue has been acid-etched away.**

Where are the lungs?

The lungs occupy the chest, on both sides of the heart. They don't go right down to the bottom of the ribcage but they do stick up a couple of centimetres above the top rib. Doctors putting needles into patients' necks have to be extremely careful not to puncture the lining of the lung. In fact, doctors' guidelines now suggest that they should use ultrasound when they put a needle into the large vein of the neck (the internal jugular vein), so that they can see exactly where the lung – and their needle – is. Below the lungs, the diaphragm seals them off from the intestines in the abdomen. The diaphragm is attached around the lower margins of the ribcage, but from there it domes up into the chest – the top of it is quite high, level with the nipples on a man's chest.

What do they look like?

The surface of adult lungs, just about anywhere in the world, is bluish-grey, but babies' lungs are a blushing, delicate pink. As a young doctor, just out of medical school, I worked in paediatric surgery (mostly helping to hold things out of the way), in amazing operations that saved the lives of tiny babies. I feel very privileged to have been part of that and I have huge admiration for paediatric surgeons, who perform the most precise of operations on the tiniest of human beings. One of the operations was to mend congenital hernias in the diaphragm: closing a hole in the division between abdomen and chest that had allowed the intestines to move up and squash the lungs. I was quite taken aback by the pristine, unpolluted, pinkness of those little lungs. But spend a lifetime living in the modern world, breathing in dirt with every breath – even without deliberately inhaling smoke – and those naturally pink lungs gradually turn an industrial grey.

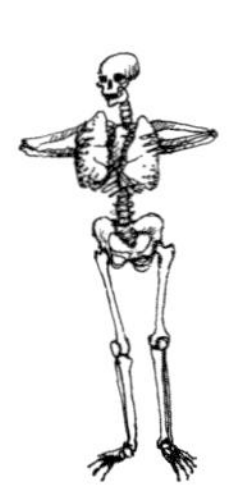

How your lungs work

Each lung is wrapped in a double bag, the pleural membrane (or just plain pleura). The pleura is a lubricated lining on the inside of the ribs and the outside of the lungs. It holds a very thin layer of fluid, just enough to make a seal between the outside of the lungs and the adjacent ribcage and diaphragm. On the inner surface of each lung, where the bronchus enters, the pleura folds back on itself, making a sealed bag.

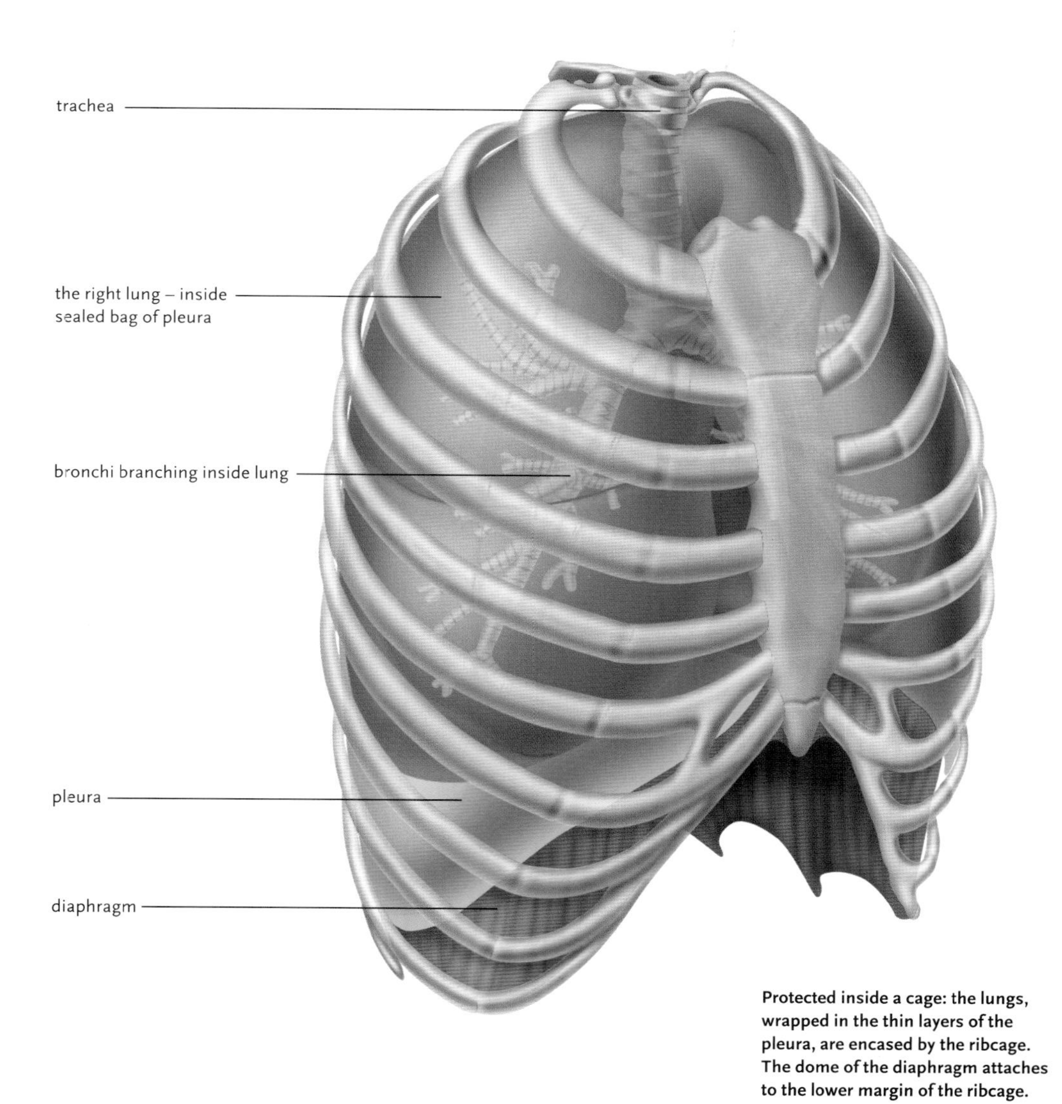

Protected inside a cage: the lungs, wrapped in the thin layers of the pleura, are encased by the ribcage. The dome of the diaphragm attaches to the lower margin of the ribcage.

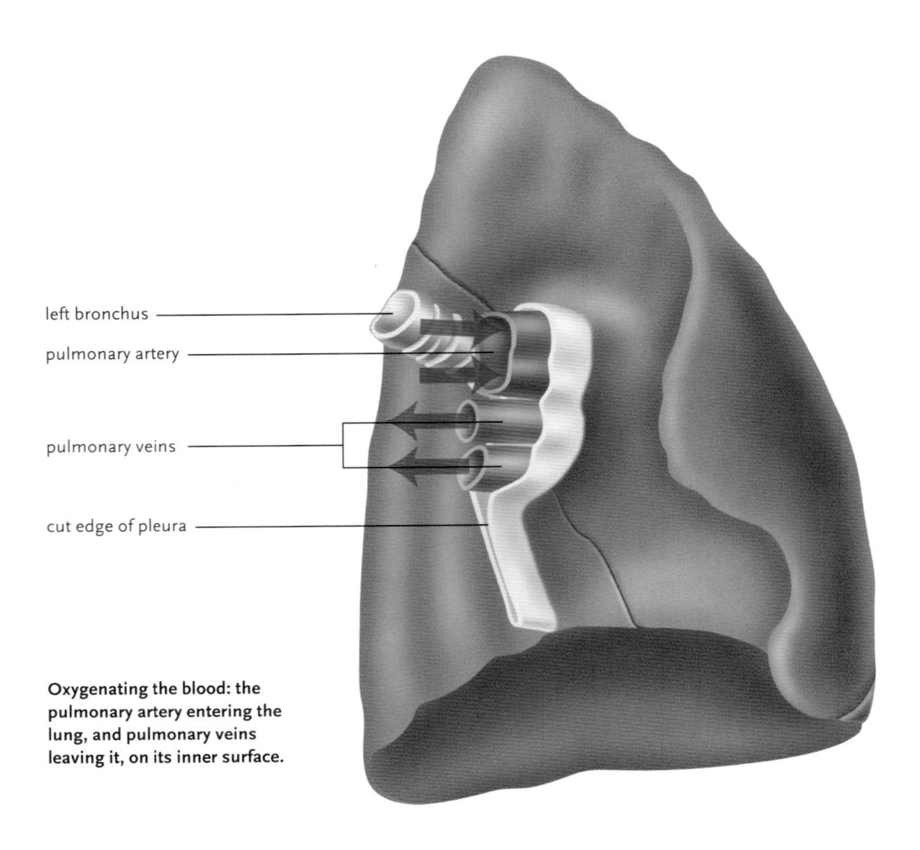

Oxygenating the blood: the pulmonary artery entering the lung, and pulmonary veins leaving it, on its inner surface.

Breathing in and out

When you take a breath in, the diaphragm contracts and flattens, and the ribs move up and out. The fluid seal inside the pleura effectively sticks the lungs to the diaphragm and ribs, so at the same time, the lungs are pulled outwards in all directions – they expand. This means the volume of each lung increases, decreasing the pressure inside, and air rushes in from the outside to fill the gap. To breathe out, all you do is let everything relax: the elastic lungs pull back, the diaphragm relaxes and domes up again and the increase in pressure pushes the air back out. Luckily you don't have to think about any of this (normally), although the muscles involved are 'voluntary' muscles – and you can exert your will over them, if you choose to: if you take a sharp intake of breath, you're deliberately flattening your diaphragm and moving your ribs. But most of the time, the breathing muscles work to a rhythm set by the lower brainstem. (This part of the brain, just above the spinal cord and

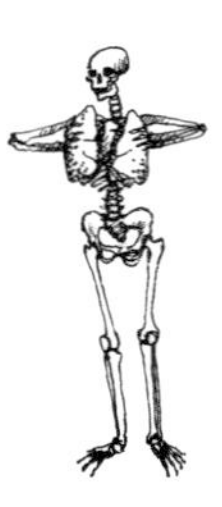

right at the bottom of your skull, controls many of the automatic processes of the body.) This is great: it means that you don't need to think to yourself: 'breathe in . . . and relax and breathe out . . . and breathe in again . . . and relax again . . .' The diaphragm and rib muscles just get on with it, without you having to spare a thought, at a gentle resting rate of about twelve to 20 breaths per minute in an adult. Elderly adults tend to breathe more quickly, taking up to 25 breaths per minute and babies breathe even faster, at some 20 to 40 breaths per minute. However, if you have a problem with your lungs, like asthma or bronchitis, you become very aware of breathing and get consciously involved in expanding and contracting your lungs. But more of that later.

A journey down the airways

Living lungs have a very strange consistency. Lightly built, with flexible tissues and full of millions of small pockets of air, they have a whipped, blancmangey texture. Lungs might be air bags but they're not entirely hollow. Each lung is a branching tree of tubes, getting smaller and smaller like the twigs on a real tree, until the very last, tiny tubes open into a cluster of tiny air sacs – the alveoli – where the business of gas exchange goes on. The trunk of the tree is the trachea or windpipe. You can feel it in your neck; it's very superficial and just a few millimetres beneath the surface of the skin.

The trachea is about twelve centimetres long, running from your voice-box or larynx to a branching point in the chest. It branches into two bronchi, one going to the inner surface of each lung. If you could take a lung out of a chest and look at this inner surface, you would see the bronchus and the blood vessels of the lung: the pulmonary artery, which brings deoxygenated blood to the lung and the pulmonary veins, which take the newly oxygenated blood away from the lung back to the heart, ready to be pumped around the body.

Once the right or left bronchus enters its lung, it branches into smaller and smaller bronchi, which go to all the lobes of the lung. For some mysterious reason, the left lung has two lobes and the right has three. These aren't just internal divisions; they are marked by deep fissures on the outside surface of the lung.

The trachea and bronchi are tubes made of muscle, kept wide open by rings of cartilage. Unlike the voluntary muscle of the diaphragm and that stretching between the ribs, the muscle inside the lungs is 'involuntary' or smooth muscle. This is very different from voluntary muscle: it looks different under the microscope and is supplied by a different type of nerve. Smooth muscle gets its nerves from the autonomic nervous system – the bit of the nervous system that looks after the automatic functions of the body. Just as in everyday use, autonomy means 'self-governing', a good term for the part of the nervous system that enables the body to control itself without us having to be consciously aware of it.

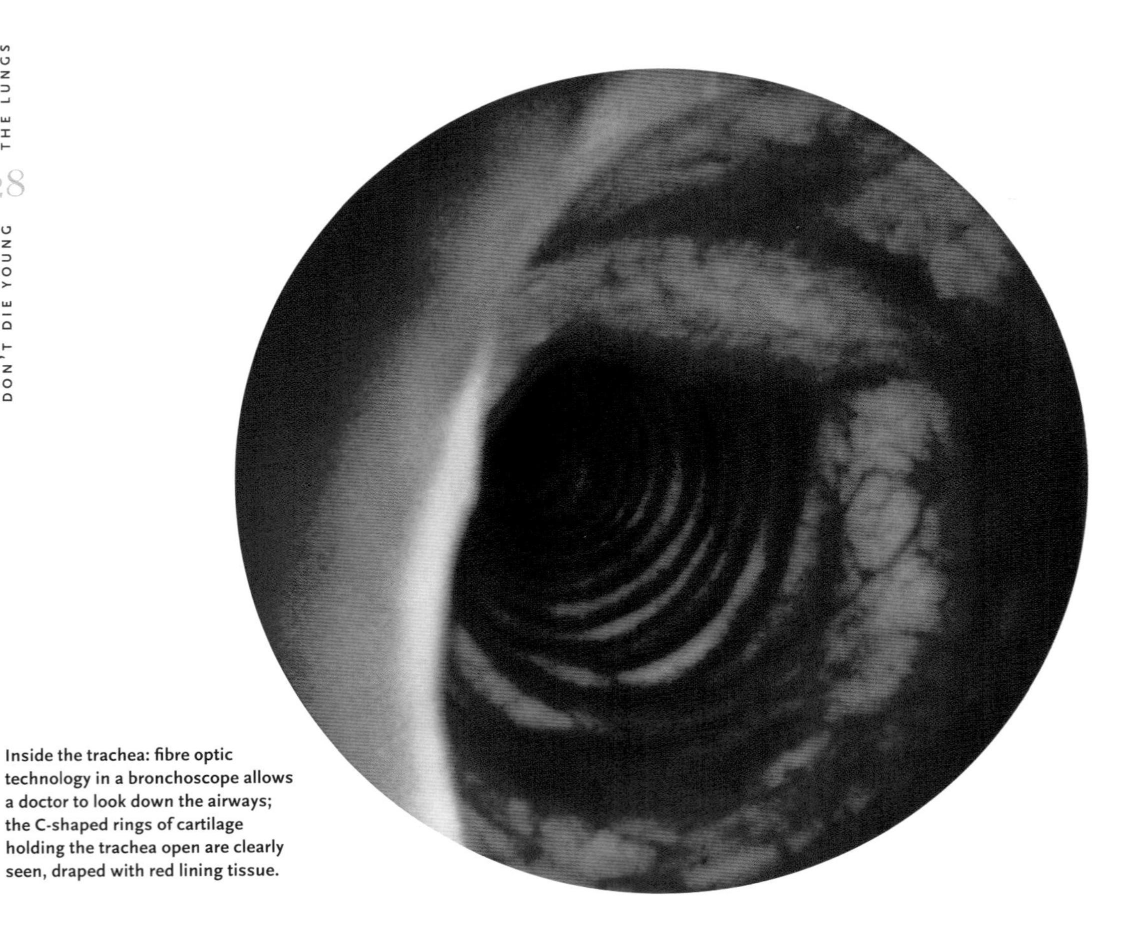

Inside the trachea: fibre optic technology in a bronchoscope allows a doctor to look down the airways; the C-shaped rings of cartilage holding the trachea open are clearly seen, draped with red lining tissue.

Within the lungs, the bronchi branch into smaller and smaller divisions, into tiny bronchioles, which finish as small clusters (like miniature bunches of grapes) of the air-filled alveoli. The smaller bronchioles have no cartilage; they're just muscular tubes. The smooth muscle in the walls of the bronchi and bronchioles means that they can get narrower or wider, to balance the pressures in the airways as you breathe. Their width increases as you breathe in and becomes narrower as you breathe out.

The airflow in and out of the lungs depends on the difference between the atmospheric pressure outside the body and the pressure inside the alveoli: if the pressure in the alveoli is lower than atmospheric pressure, air will rush in; if the air inside the alveoli is at a higher pressure than atmospheric pressure, it will be forced out. But on its way out, the air gradually loses pressure as it moves up through the airways. To keep the pressure up enough to get the air out, the airways narrow themselves as you breathe out.

As well as changes with each breath, there's a daily rhythm in the width of the airways, and they are at their narrowest in the early morning, at about 4 a.m. The constriction of the airways in an asthma attack is an exaggerated version of this response – and because it's superimposed on the daily rhythm, asthma symptoms tend to be worse in the early morning. Various substances that you breathe in can also have an effect: the airways may narrow in response to cold air, dust and cigarette smoke, for example.

How the lungs keep themselves clean

The lungs have a fantastic self-cleaning device to rid themselves of all the dirty particles that come into them with each breath. Every branch of every airway is lined with a very thin film of mucus, which traps dust particles, bacteria, pollen – anything. The mucus is made by 'goblet cells' scattered in the lining of the bronchi and bronchioles. They really do look like little goblets, with their cups constantly overflowing as they produce more and more mucus. Where does all that mucus go? Well, the

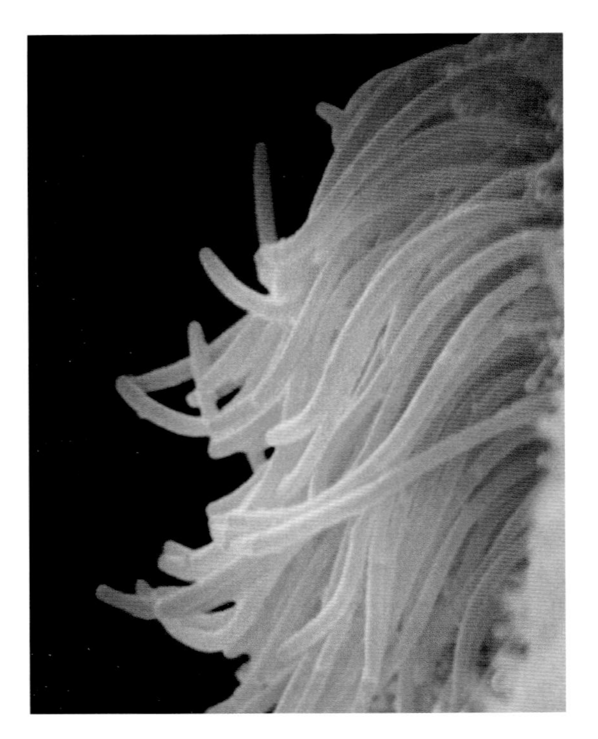

A magic carpet: cilia seen under an electron microscope.

other cells lining the airways have a strange surface that looks like velvet, under high magnification, because the cells are carpeted in a layer of tiny hair-like projections called cilia – and these cilia *move*. Each one is absolutely minute but they all move in a coordinated way to waft the mucus along. They waft it up the bronchioles, up the bronchi and out of the lungs, up the trachea, and through the larynx. And then you swallow it. In medical terminology, this moving layer of mucus propelled by the waving of the tiny cilia has a fantastically redolent name: the 'muco-ciliary escalator'.

Oxygenating the blood

Looking at the alveoli (the tiny bunches of grapes at the end of the smallest bronchioles) under the microscope, you can see how the lungs do their job of oxygenating the blood. Each alveolus (Latin for 'small cavity') is a tiny little pocket of air. The lining of each alveolus is just one cell thick – and those cells are flattened right down. Outside each alveolus is a network of capillaries; tiny blood vessels whose walls are also just one flat cell thick. The thinness of these membranes is crucial to the function of the lungs. Oxygen has to pass across the wall of the alveolus and across the wall of the capillaries to get into the blood – so the thinner those walls, the better. However, the thinness of the alveolar walls presents a potential problem. As you breathe in, the air pressure inside the alveoli is lower than atmospheric pressure; at such low pressures, the thin-walled alveoli are in danger of collapsing in on themselves. The cells lining the alveoli overcome this problem by making a special fluid called surfactant, which reduces the surface tension around the inner surface of each alveolus and helps it to stay open.

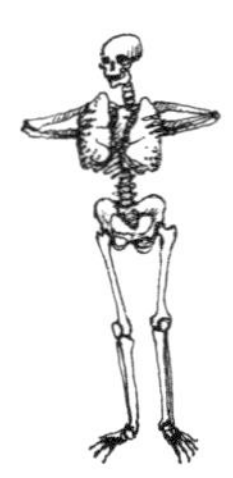

Premature babies often have problems breathing; sometimes their alveolar cells haven't quite flattened, so they're more of a barrier to oxygen. On top of that, their alveolar cells may not have started to make surfactant (you don't need surfactant until you start breathing air so the foetus has no need for the stuff while in its mother's womb). Being born prematurely takes a baby's lungs by surprise: breathing air was not something it expected to do until a good nine months after conception. Without surfactant, the alveoli can collapse and breathing becomes very, very hard for the baby. Babies with this problem can be helped by being given pressurized air to breathe and by squirting an artificial surfactant into the lungs – but it's never quite as good as the real thing.

Returning to the journey of the oxygen: once in the blood, it attaches itself to molecules of haemoglobin, which are found in the doughnut-shaped red blood cells. Haemoglobin is a molecule specifically designed for carrying oxygen. However, the dangerous gas carbon monoxide is capable of attaching itself much more strongly to haemoglobin than oxygen. This is what makes carbon monoxide such a lethal airborne pollutant: it takes up room on the haemoglobin molecules and lowers the amount of oxygen the blood can carry. Many city centres have pollution sensors, monitoring carbon monoxide levels on the street. This technology is also breaking out into mobile devices, with portable sensors used to measure carbon monoxide levels on pedestrian and cycling routes.

As well as picking up oxygen, as the blood passes through the capillaries around the alveolus, carbon dioxide crosses the capillary wall, into the air inside the alveoli, ready to be breathed out. The 300 million alveoli in each lung provide a huge area for gas exchange (getting oxygen into blood and getting carbon dioxide out). The blood is a transport system, carrying lorry-loads (or rather, red-blood-cell-loads) of oxygen to the tissues of the body and bringing their carbon dioxide back. The capacity of your lungs can be increased with regular exercise, making gas exchange more efficient. The more often you exercise – and it doesn't matter what type of exercise – the better your lung capacity.

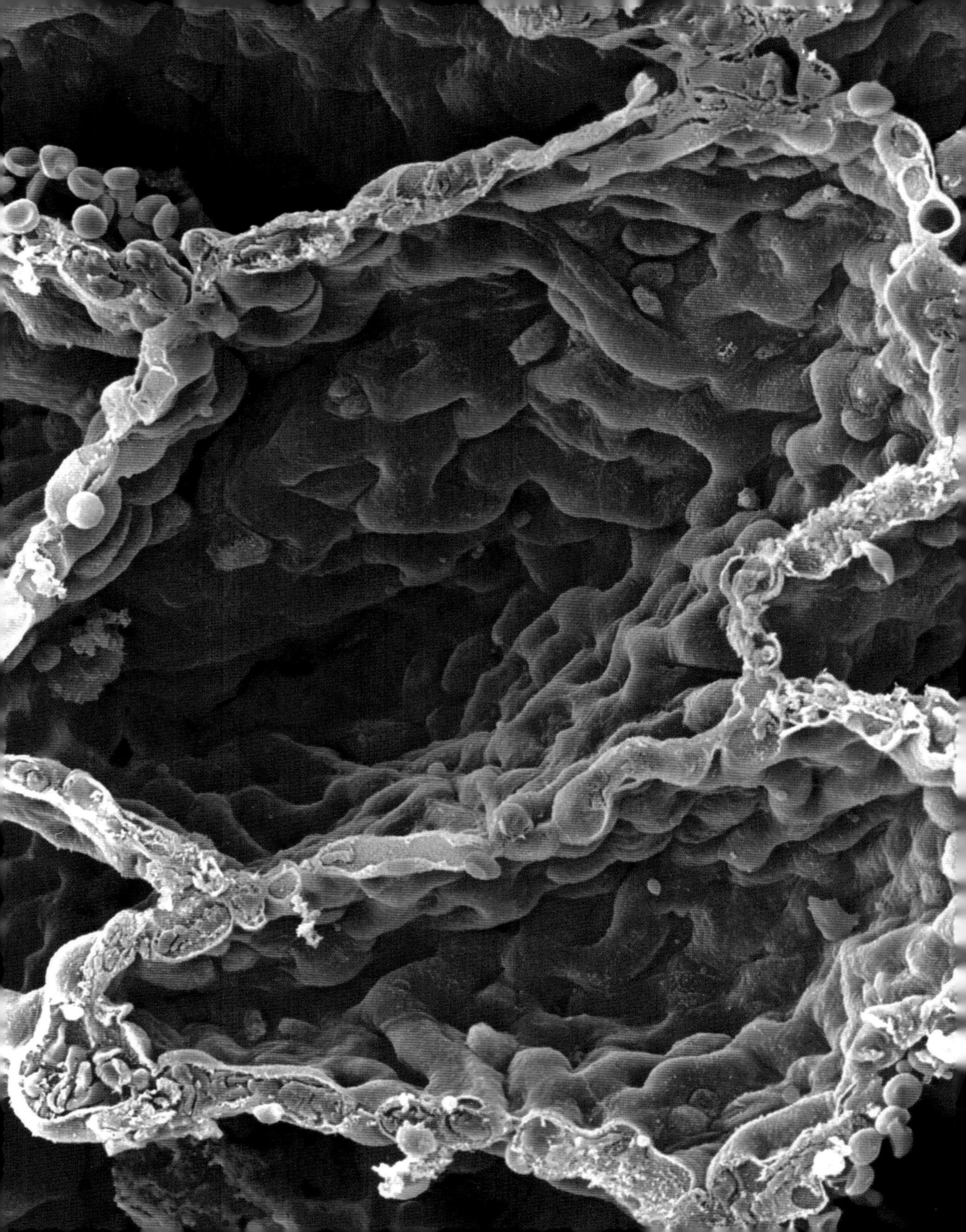

Common lung problems, and how to prevent them

So much for healthily functioning lungs. Now we turn to some common lung problems; especially those where life and diet play a significant role. These are problems you have some control over, and may even be able to prevent by living healthily. They are also problems that have a huge impact on the health of the nation. I'll look at smoking and its consequences for the lungs and then in detail at the rise of asthma.

Smoking, lung cancer and COPD

The link between lung cancer and smoking is very well known. There is no way around it: smoking is *the* most important factor causing lung cancer (and it also causes others, such as cancer of the mouth, oesophagus and bladder) and it massively increases the risk of heart disease and strokes. Cigarette smoke is full of toxins and cancer-causing compounds. These dangerous compounds cause mutations in genes (i.e. alterations in the DNA code), which cause cells to become cancerous. It doesn't have to be your own smoke doing the damage: second-hand smoke can lead to lung cancer too. Outdoor air pollution also contributes; about eleven per cent of lung cancers in Europe can be blamed on pollution.

Smoking also causes Chronic Obstructive Pulmonary Disease (COPD), which used to be known as chronic bronchitis. COPD is predicted to become the third most common cause of death by 2020. More than nine in ten cases are caused by smoking. In COPD, the lungs produce too much mucus, the bronchial walls become scarred and the airways get narrower. Getting breath in and out of the lungs becomes an effort. The alveoli collapse together making larger air sacs, which aren't as good at gas exchange: a condition called emphysema. The fundamental role of the lungs is disturbed: without the large surface area in the millions of tiny alveoli, the lungs become worse and worse at the

Left **The air sacs: a slice across two alveoli showing their extremely thin lining, just one cell thick. Doughnut-shaped red blood cells spill out of capillaries at the edges of the alveoli.**

all-important exchange of oxygen and carbon dioxide between blood and air. Eventually, COPD can cause fatal 'respiratory failure', when the lungs are no longer up to the job of getting enough oxygen into the body and enough carbon dioxide out.

Asthma

In the UK in the twentieth century, there seemed to be a huge rise in the level of a devastating, life-threatening lung disease. In this case, I'm not talking about lung cancer. So you may think I'm being a little over the top about a disease we can control with little blue aerosol inhalers. But asthma is a huge health problem, affecting about 300 million people across the world. In the UK, well over five million people are asthmatic: around one in twelve adults and one in eight children. In terms of their overall health and how much people's daily lives are affected, asthma is a more important long-term health problem than diabetes. And asthma is a killer – at least a thousand people die from asthma every single year in the UK.

The word 'asthma' comes from Greek and means 'to breathe hard'; anyone who has this disease will know and dread the feeling of tightness in the chest. Breathing stops being something you do without thinking about it. The struggle to breathe takes over your body and mind – you have to really concentrate on expanding your lungs as much as you can, and fight to draw each breath in and to breathe it out. The bronchioles become inflamed and narrow, and it's that narrowing which means it's harder to get the air in and out of the lungs. It's like a three-lane motorway being reduced to two lanes for roadworks, but with the same volume of traffic struggling to get through.

The narrowing of the bronchioles creates a strange effect; as the air whistles through the narrowed tubes, the asthma sufferer wheezes. I remember listening to my own wheezes, as a five-year-old, long before I'd heard of asthma; lying in bed under my feather duvet (which I was allergic to), wondering at the gentle symphony of flute-like sounds coming from my chest.

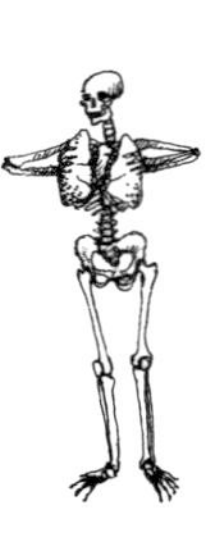

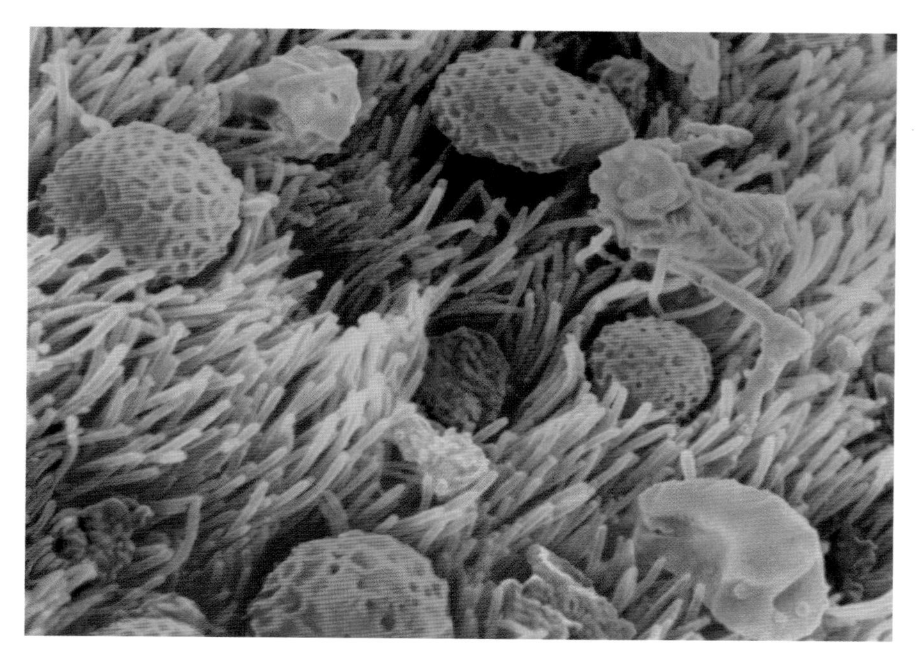

Pollen grains (pink) and dust particles (blue) scattered on the velvety, ciliated lining of the trachea. The cilia will waft these particles – stuck in mucus – up and out of the lungs.

Allergens: An asthmatic person's airways are 'hypersensitive'; they over-react to things – 'allergens' – in the environment, such as pollen, dust mites and animal fur, that don't cause other people any problems at all. They are 'allergic' to these substances: the body is reacting to something it recognizes as 'other'. This recognition is carried out by the immune system. This system constantly checks to make sure that there is no 'other' in our bodies; that all the stuff inside us is just us. So, when nasty outsiders like viruses and bacteria do manage to get into the body, they are quickly recognized as other and the immune system mobilizes its defence mechanisms.

What happens in an allergic response is that the body makes a mistake, over-reacting to not particularly harmful substances. To use a politically charged analogy, the body launches a pre-emptive strike against something that doesn't actually have any weapons.

Asthma is one of a set of allergic diseases where the body's reaction may happen away from the initial point of contact with the allergen.

This type of reaction is called 'atopic'. It happens because antibodies, defensive proteins produced by white blood cells, travel round the body and trigger reactions in areas nowhere near the place that came into contact with the allergen. Atopic diseases include the skin rashes of eczema, the weeping eyes and dripping noses of hay-fever and the narrowed bronchioles and wheezing of asthma. These three are linked by an underlying predisposition to hypersensitivity. If you've had eczema, asthma or hay-fever, you're more likely to develop the other two than people who've never had an atopic condition. If you've had all three, you've scored the atopic hat-trick.

In modern western societies, the rise in the number of people with asthma is part of a general rise in the number of people with atopic diseases. Hay-fever was first described in the early nineteenth century, when it was rare. Today, one in two people in developed countries is allergic to something or other in their environment. The rate of allergic disease in Britain is ten to 15 times higher than it is in Asia. The pattern of asthma in Germany, before unification, showed that this must be something to do with lifestyle rather than genes: affluent West Germany had a higher rate of asthma than less-privileged East Germany (which is now catching up). The increasing rate of allergies in the population seems to be linked to western lifestyles – but how?

Asthma, like many diseases, is multifactorial. This means that many factors conspire to produce the condition that we recognize as asthma. It also means that any increase in the prevalence of asthma is probably due to several factors and it's very difficult to work out which are the most important. There are both genetic and environmental factors at work; and a bewildering array of studies link asthma to all sorts of potential causes, including air pollution, various foods, and chemicals we might be breathing in or eating.

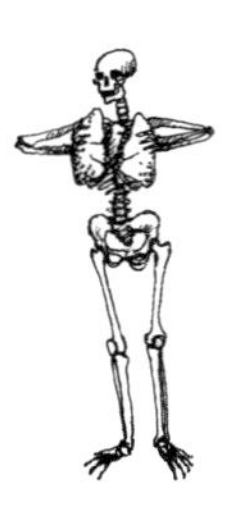

Hygiene: Atopic diseases seem to develop early in life and the 'hygiene hypothesis' puts this increase down to a lack of exposure to environmental allergens, like bacteria, during childhood. We are just too clean! There is

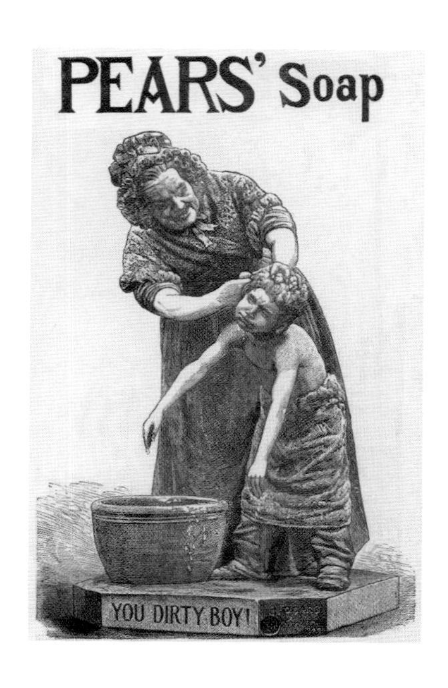

The Victorians may have put cleanliness next to godliness, but are our children now *too* clean?

absolutely no doubt that the 'hygiene revolution' of the nineteenth century had a huge impact on life expectancy when it was recognized that tiny 'germs', invisible to the naked eye, were responsible for many diseases. Infection had been the Number One killer of humans for thousands and thousands of years. During this period, we worked out how to reduce the risk of getting infected, simply through good hygiene and clean water. The Victorians put this into practice with quasi-religious zeal – carbolic soap became the saviour of humanity. The revolution in hygiene was probably as (if not more) important as the discovery of antibiotics in terms of its impact on reducing the prevalence of infections. But it seems that we have gone a bit too far – our human bodies have evolved in an environment seething with bacteria; we don't seem to be able to cope if we get rid of them completely. Attempting to remove all the bacteria and viruses from our immediate environment may be doing us more harm than good.

The problem starts in early childhood because that's when the immune system is learning to recognize things which might cause an infection. A baby's immune system is immature and reacts to potential threats in a generalized way. But as the immune system 'sees' more bacteria and viruses, the response becomes honed to a more precise and targeted attack and the generalized response is turned off. If the immune system is shielded from exposure to bugs, the targeted response doesn't develop and the generalized response stays around, causing atopy,

The enemy of asthmatics: the humble dust mite chomps away on the dead skin we cast off; there are millions of these mites in every house.

allergies and asthma. A baby that spends time with lots of other children, in a big family or in a nursery, or who grows up close to animals, and actually gets infections (especially viral infections like colds, measles and chicken pox) is less likely to develop atopic disease. But a child in a small family, with 'high' standards of hygiene and who is given antibiotics early in life, is more likely to get asthma, eczema and hay-fever. We need to strike a balance between protecting against serious and life-threatening diseases whilst allowing children to be exposed to some bugs that help their immune systems mature.

Early contact with allergens does seem to be very important – a little dirt is good for you. Researchers are testing a 'vaccine' made of harmless soil bacteria to see if this reduces the development of allergies. It also seems that probiotics (or 'friendly bacteria') might have a role to play in reducing atopy even before we are born. A study published in *The Lancet* showed that babies born to mothers who took probiotics before and after birth were half as likely to have eczema.

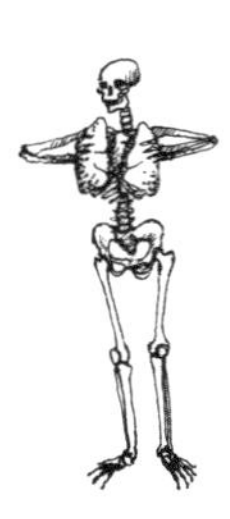

The effects of breastfeeding on atopy are somewhat controversial; for a long time, breastfeeding was considered to protect a child from developing allergic diseases but some recent studies have suggested that it is linked to an increase in asthma and atopy. However, the results aren't conclusive enough to suggest that babies should be denied the known benefits of breastfeeding.

For people who have developed asthma or other allergic conditions, avoiding allergens is definitely recommended. House dust mites, a common allergen, are tiny creatures about a third of a millimetre long that live all around us in our houses. About two million house dust mites can live happily in an average mattress. The scientific name of the European house dust mite, *Dermatophagoides*, comes from the Greek for 'skin-eating'; we each shed about a gram of skin every day, which forms the staple diet of these tiny bugs. House dust mite levels have increased over the last few decades. They particularly like carpets, especially deep-pile woollen carpets, and they thrive in modern centrally heated and double-glazed houses. Airing the house is recommended, along with regular dusting – this pains me, because I hate housework and dusting sets my asthma off, but using a damp duster can at least stop the dust getting into the air. Cutting down exposure to allergens like house dust mites has been shown to reduce the symptoms of asthma.

Diet and exercise: Some scientists link lifestyle changes in developed countries to the increased rate of asthma. Increasing affluence seems, paradoxically, to be linked to an unhealthy change in diet: fewer fruit and vegetables (which contain valuable antioxidants), more margarine, and less butter and oily fish. Could diet, physical activity or obesity be related to asthma? Taking certain vitamins and oily fish or taking omega-3 supplements have been suggested as ways of reducing the risk of asthma or of controlling asthma symptoms but the results of scientific studies have largely been disappointing. However, just because more research needs to be done on diet and asthma, don't take this as an excuse not to eat healthily! Obesity is a risk factor for asthma and COPD. As well as imposing a mechanical restriction on lung function, obesity might somehow predispose a person to airway hypersensitivity.

More and more studies point to physical exercise as an important consideration for asthmatics. This may not come as a surprise: most people know that asthmatics often get wheezy with physical exertion. But what isn't so widely appreciated is the role of exercise in preventing asthma. Regular physical exercise has a significant effect on reducing the impact of asthma and improving the quality of life of asthma sufferers. American guidelines for people with asthma include them getting regular exercise but the message seems to be rather slow reaching this side of the Atlantic. Exercise gets a quick mention in the British Thoracic Society's guidelines for asthma management, but rather bizarrely, it comes under 'complementary and alternative medicine' and is seventh in the list, after herbal medicine, acupuncture, air ionizers, homeopathy, hypnosis and massage. As well as improving the health of people who already have asthma, it is possible that physical exercise in childhood could protect against asthma developing in the first place. The decrease in physical activity amongst children in developed countries may be an important factor in the rising rates of asthma.

What you can do: Some researchers in this area talk about an 'allergy community', putting their emphasis on medical therapy and allergen avoidance. That certainly tallies with my own early experience of being an asthma sufferer. It started with my first encounter with an allergy specialist, who placed known allergens on scratches along my arm and deduced, from the developing weals, that I was allergic to – well, just about everything – cats, dogs, grass pollen, house dust and house dust mites ... I imagined a future living in some kind of NASA-like sealed tent! Asthma was presented to me as something that I had; something that I myself could do very little in a positive way to control. It was all about avoiding things (like house dust, pollen and exercise) and taking drugs; about being utterly dependent on a small, blue aerosol pump. As a seven-year-old who had never really enjoyed PE, asthma was a fantastic get-out

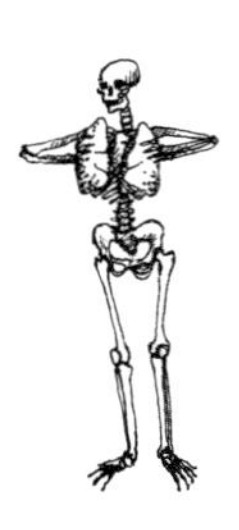

HEALTHY SCEPTICISM: OXYGEN BARS

Here's a trendy new 'therapy' designed to raise your energy levels, banish tiredness, ease away stress and give you a 'natural high'. Oxygen bars have been in America since the early 1990s – and now they're springing up in Britain. You can have your oxygen flavoured with apple, mango, strawberry, watermelon, peach, orange, cranberry, peppermint, *piña colada*, lime or chocolate. For the discerning client, there are various colours of nose hoses. Has the world gone mad?

Oxygen bars tend to be advertised in a vague, fluffy way, talking about energy levels and relaxation, but usually avoiding making any medical claims about the health benefits of breathing oxygen. And this is because there is no scientific evidence to support such a claim. If you're healthy and your lungs are working well, then you can extract all the oxygen you need from air (which is about 21 per cent oxygen). Certainly, oxygen may be given to patients in hospital, but only when the level of oxygen in their blood has become dangerously low. There's a chance that trying to 'top up' could be even be bad for you. Hospital patients given too much oxygen have been shown to suffer oxidative stress (see page 240). Prolonged exposure to high levels of oxygen is very bad for you indeed, causing a slow heart, dizziness, hallucinations and loss of consciousness (as opposed to the 'energy boost' that the oxygen bar ads claim a ten-minute shot of oxygen could confer). The aromas added to the oxygen could also be a problem, potentially irritating the lungs. Although there's no hard evidence that taking a short blast of an oxygen bar may be harmful to a healthy person (apart from, of course, producing severe pains in the wallet area), it could be dangerous for people with heart and lung diseases. I've tried it and didn't feel any effects at all, other than that the nose hose was a little uncomfortable and it smelt a bit strange.

clause. I'm only grateful that the exercise I may have missed out in school was more than made up for by taking my dog for a walk and climbing trees after school. I was very lucky to grow up on the edge of a big wooded park and having a dog to look after meant I was in there every day. I spent much more time playing in the woods, getting muddy in the stream and being outdoors than I did watching television inside.

As an adult with asthma, exercise has certainly been mentioned in my regular check-ups but the real emphasis is on whether I'm being 'good' and taking my steroid inhaler regularly. Again, I've been ushered into a passive role – keep avoiding the allergens and keep taking the drugs.

Of course, I'm not suggesting that asthmatics should throw away their inhalers or seek out allergens. That would be nonsensical and irresponsible. Avoiding things which set off your asthma is common sense and inhalers really help reduce the inflammatory changes in your lungs, as well as providing instant and welcome relief from wheezing. In a full-blown asthma attack, inhalers could save your life. But if exercise is so important in improving the fitness of your lungs and reducing the effect of asthma, it surely would make sense to make it a central pillar of asthma management. It's a way of taking back control – a really positive way to improve health.

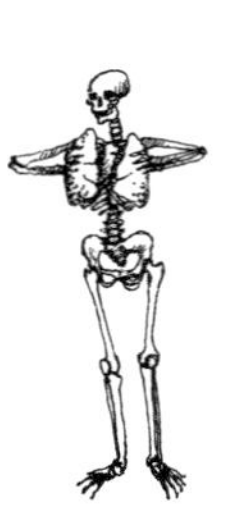

Five ways to keep your lungs healthy:

- Don't smoke – cigarette smoking increases your risk of lung cancer, bronchitis and COPD.
- Exercise regularly, eat a healthy diet and keep your weight down, in order to reduce your risk of developing asthma and COPD.
- Cut down exposure to biological pollutants in the house (for example, mould spores and house dust mites) by keeping the house clean and well aired.
- Avoid indoor chemical pollutants like air fresheners, perfumed candles, fly spray and cigarette smoke.
- Cut down exposure to outdoor pollutants, such as car exhaust fumes and smoke.

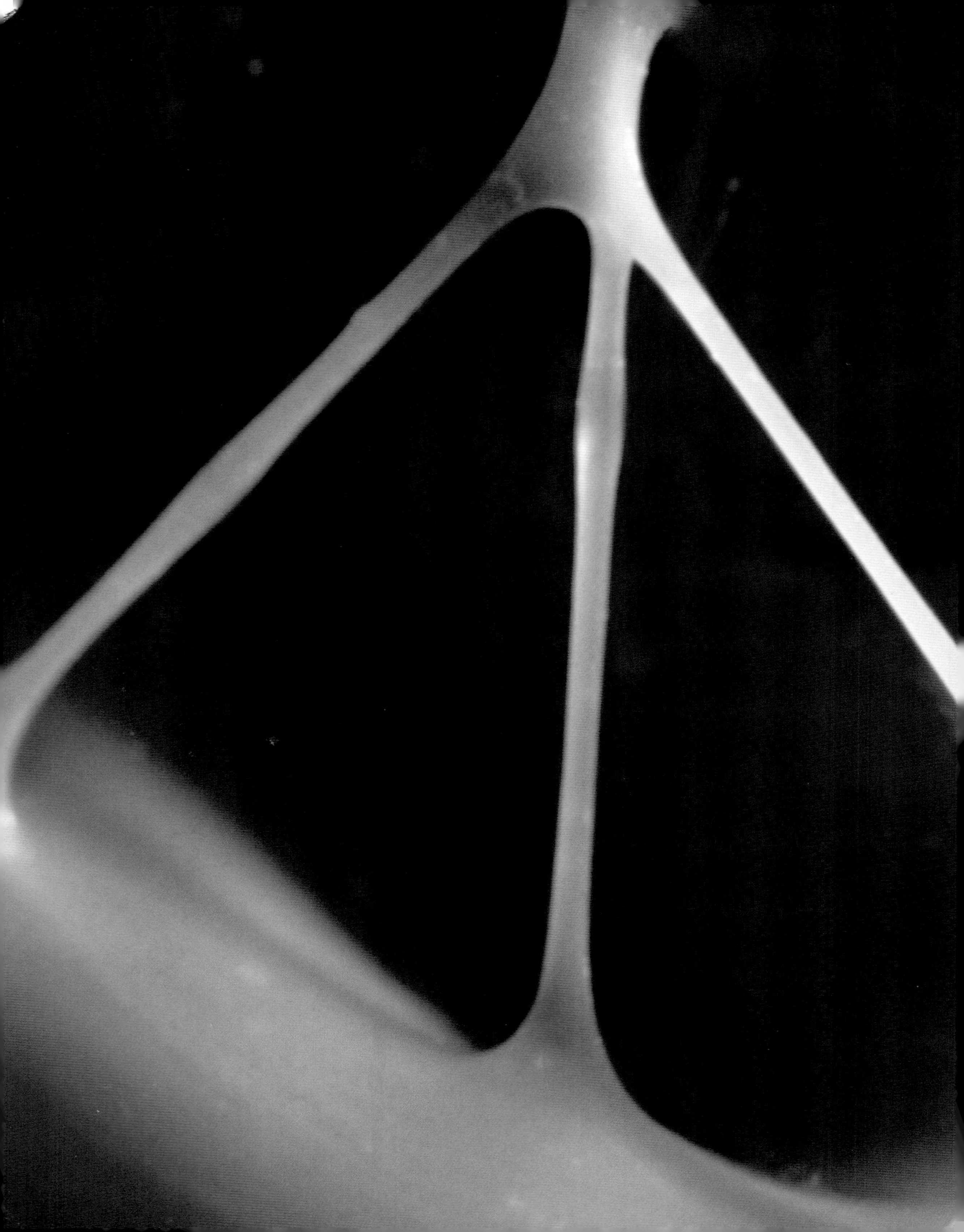

TWO

THE HEART

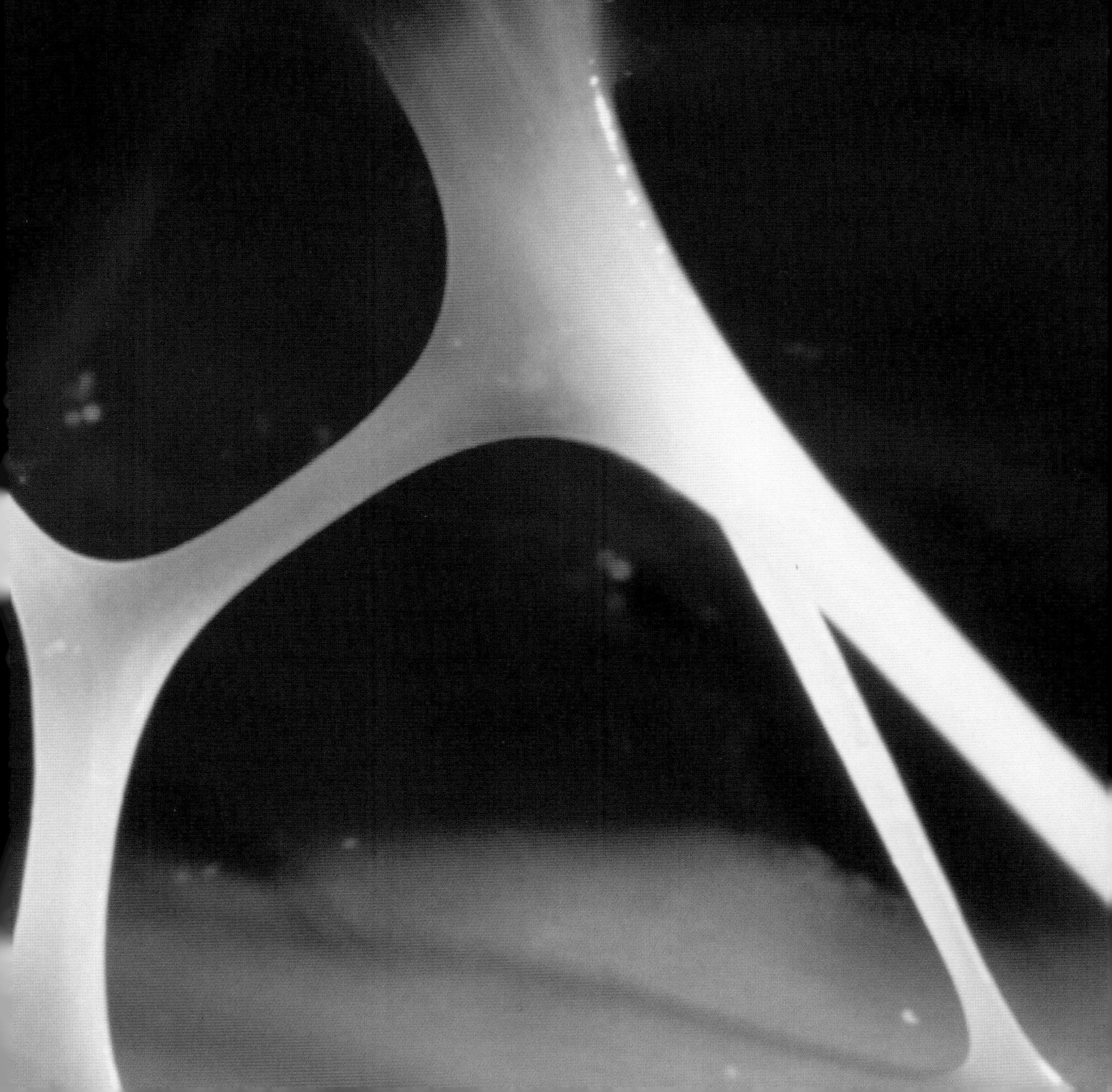

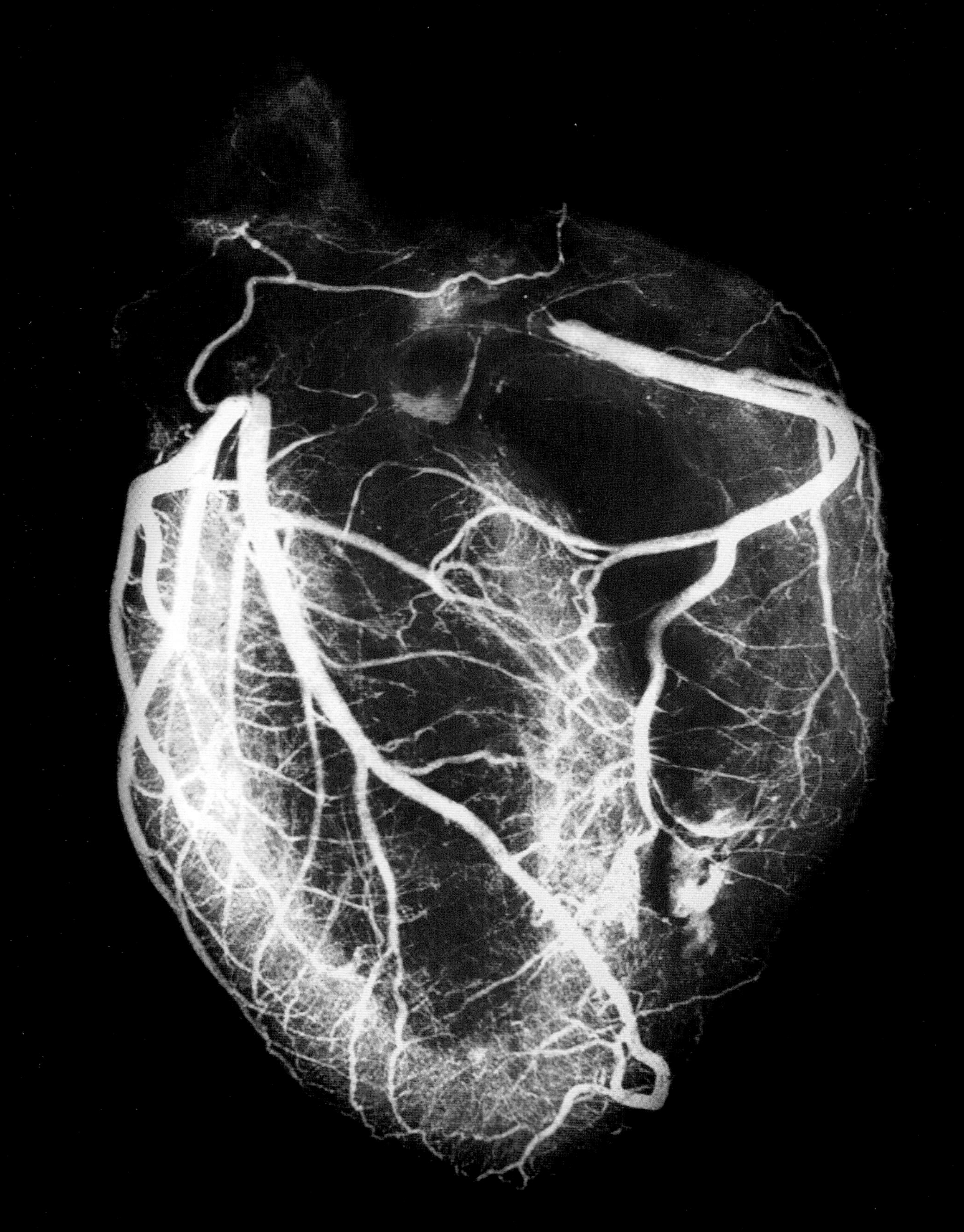

It's all very well drawing air into the lungs and exchanging gases with the blood, unloading carbon dioxide and loading up with oxygen. But that blood has to move away from the lungs and go off to the rest of the body, where the oxygen is needed. Something has to create a current in the bloodstream, and that something is the heart.

The heart pumps blood around the body to take oxygen out to distant tissues. The bloodstream is a universal delivery system, carrying nutrients that have been absorbed from the gut, white blood cells which look out for infections, and the chemical messengers of the body, hormones. However, unlike the supermarket delivering groceries, the bloodstream also takes away the waste.

The cardiovascular system (heart and blood vessels) is a complicated network of tubes through which the heart pumps the blood. Many of the problems affecting the cardiovascular system boil down to narrowing or blockages of the tubes; in very simple terms, this is what happens in angina, heart attacks, peripheral vascular disease and deep venous thrombosis.

Previous pages **Tugging on the heartstrings: visible to the naked eye, these are the tendinous cords that keep the heart valves working.**

Left **A crown of arteries: encircling the heart, the two coronary arteries supply the heart muscle with oxygen to keep it pumping.**

Where is the heart?

Hold up your two fists and place them together: that's about the size of your heart. It sits in the middle of your chest, slightly skewed over to the left. If you're a man, the apex (the pointy bottom end) of your heart lies level with your left nipple. If you're a woman, this isn't such a reliable landmark!

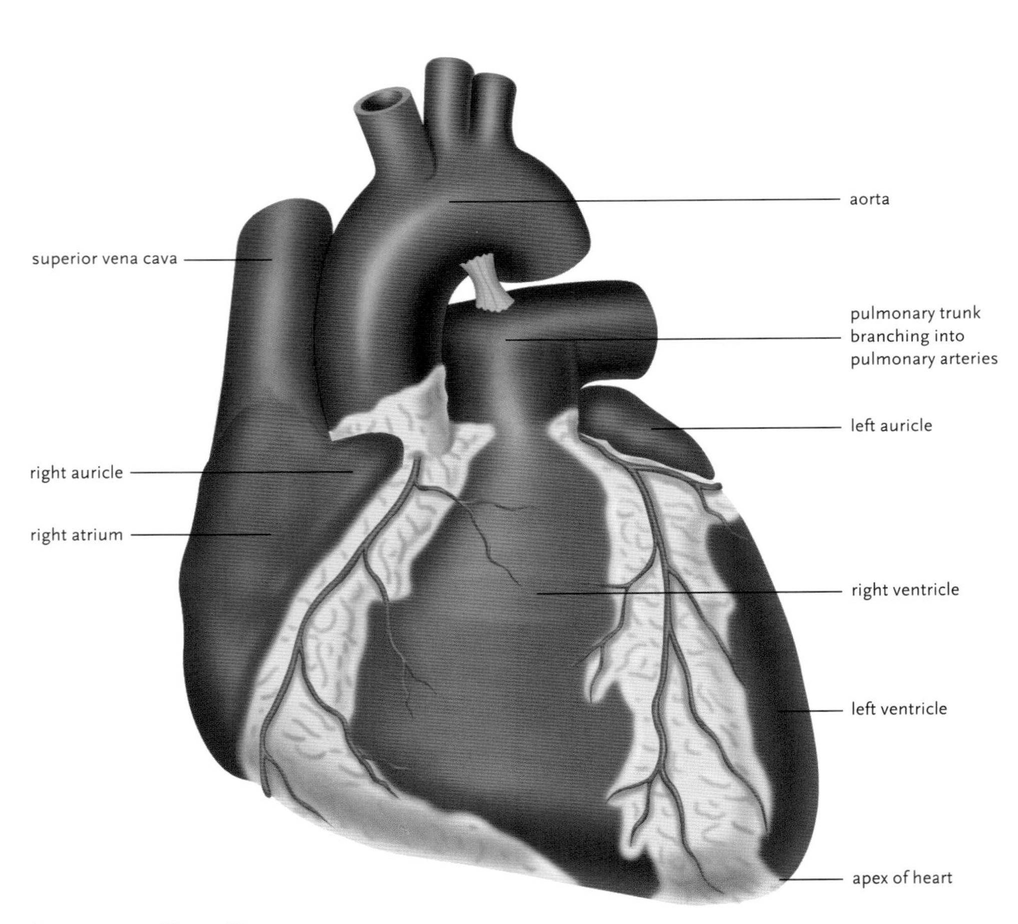

The pump: two sides working as one to pump deoxygenated blood to the lungs and oxygenated blood off to the rest of the body.

How the heart works

The heart has veins coming into it and arteries going out of it. It has four chambers: the left atrium and ventricle and the right atrium and ventricle. The blood on each side of the heart is kept completely separate: the deoxygenated blood in the right side is there to be pumped to the lungs and the oxygenated blood from the lungs returns to the left side of the heart to be pumped out to the rest of the body. Both sides work together, so the right and left atria contract at the same time, and so do the right and left ventricles.

The atria (Latin for 'entrance hall') are the chambers of the heart that first receive incoming blood from veins. They have weird little appendages which almost look like small floppy ears, and they're called auricles (Latin for 'little ear'). The insides of the atria are smooth-walled but the auricles have very strange, ridged muscle inside. It seems rather odd to have such little, blind-ended, rough-walled appendages attached to the atria – and even a bit inadvisable. They look like places where clots might form, little cul-de-sacs that invite stagnation. And indeed, although the auricles might play some kind of role as blood reservoirs, they are common sites for the formation of clots.

Inside the right atrium is an elliptical depression in the wall between it and the left atrium; this is the remains of the oval foramen (foramen is a common anatomical term for a hole). In the fetus, the oval foramen allows blood to move directly from the right to the left atrium. This is because oxygenated blood from the placenta comes into the *right* side of the fetal heart, and needs to get into the left side quickly so that it can get out to the tissues. So it takes a short-cut through the oval foramen and bypasses the lungs. It's vital that this foramen closes at birth, and it's set up like a valve, which closes when the baby takes its first breath. In some people, the valve is faulty and a 'hole in the heart' persists. If this is small, you can get away with it, but larger holes need to be surgically closed.

Each atrium opens into the ventricle through a valve – a three-flap (tricuspid) valve on the right and a two-flap (bicuspid) valve on the left. The bicuspid valve is also referred to as the mitral valve, because the two cusps together form a shape like a bishop's mitre. The edges of these valve flaps have 'guy-ropes' (called the 'tendinous cords') attached to them, which stop them turning inside out when the ventricle contracts: the heart really does have heartstrings.

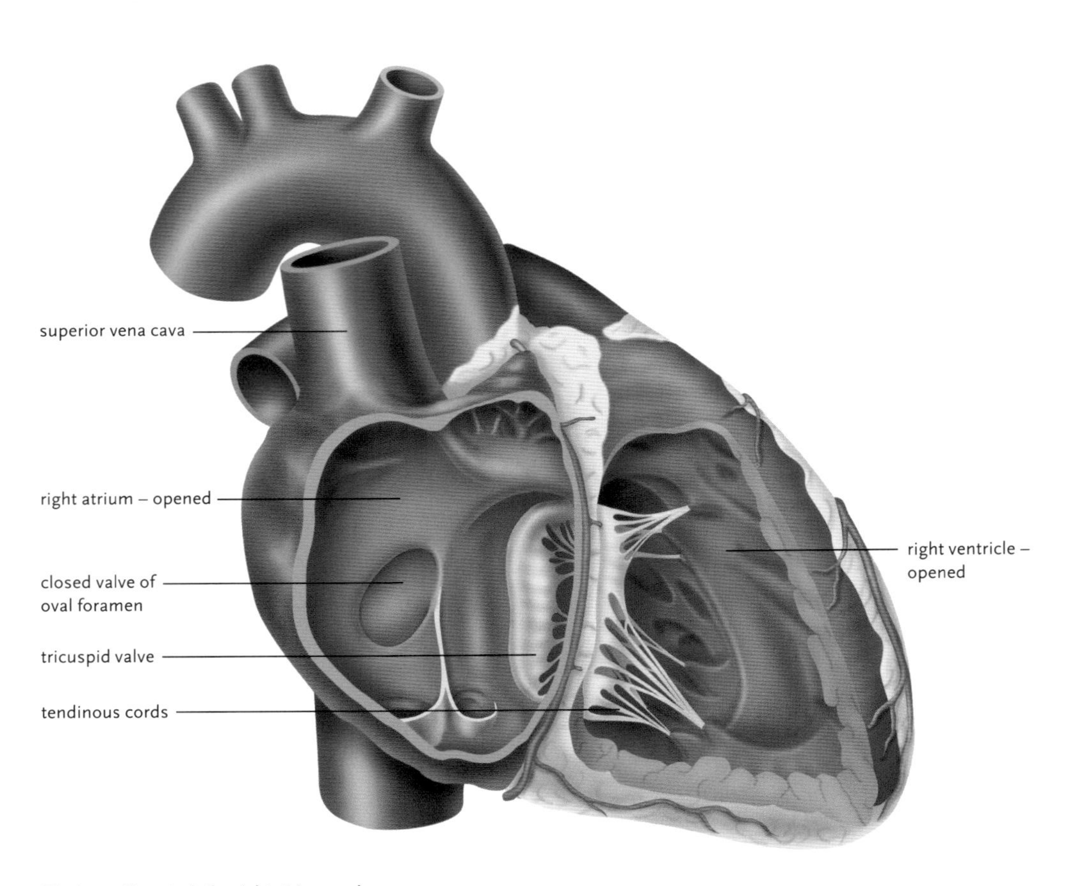

The heart dissected: the right atrium and ventricle have been 'opened'; the tricuspid valve is open like this between beats, allowing blood into the right ventricle, but then shuts as the ventricle contracts.

The word ventricle (Latin for 'small belly') is used here in the sense of the heart's ventricles being hollow structures. The ventricles have rough muscular walls. The muscle wall of the left ventricle is about three times as thick as the right's, so it produces a higher pressure when it contracts. This leads us to the reason for having separate left and right sides of the heart. The blood being pumped out to the body (from the left side of the heart) needs to be at high enough pressure to get blood all the way to the tips of your toes and fingers and up to the brain. However, if the pressure were too high in the lung's circulation (pumped from the right side of the heart), there is a danger of fluid being forced out of the capillaries into the alveoli, which would effectively drown you from the inside out. This is what happens in heart failure: the heart doesn't pump blood quickly or strongly enough and pressure builds up in the lungs. People with heart failure find it hard to breathe, because their alveoli fill up with fluid, and they have frothy sputum when they cough. It's a serious situation but it can be relieved by drugs which make the heart pump more strongly.

What makes the heart beat?

Like the lungs, the heart gets on with its work without us thinking about it; it's certainly not consciously controlled. The basic heart beat isn't controlled by the brain at all – not even subconsciously; the heart controls itself. Heart muscle is a special type of involuntary muscle. Unlike the smooth involuntary muscle, with small, spindle-shaped cells, that is found in other organs and in the walls of blood vessels, heart muscle is a network of cells.

A wave of electrical charge can spread quickly through this network, like a forest fire spreading through dry brush, and a wave of contraction follows hot on its heels. But what kicks off this wave? At the top of the right atrium, a small collection of specialized heart muscle cells automatically changes its electrical charge, recharging and discharging about 70 to 80 times per minute. Every time it discharges, a wave of contraction spreads across the atria and pumps the blood into the ventricle.

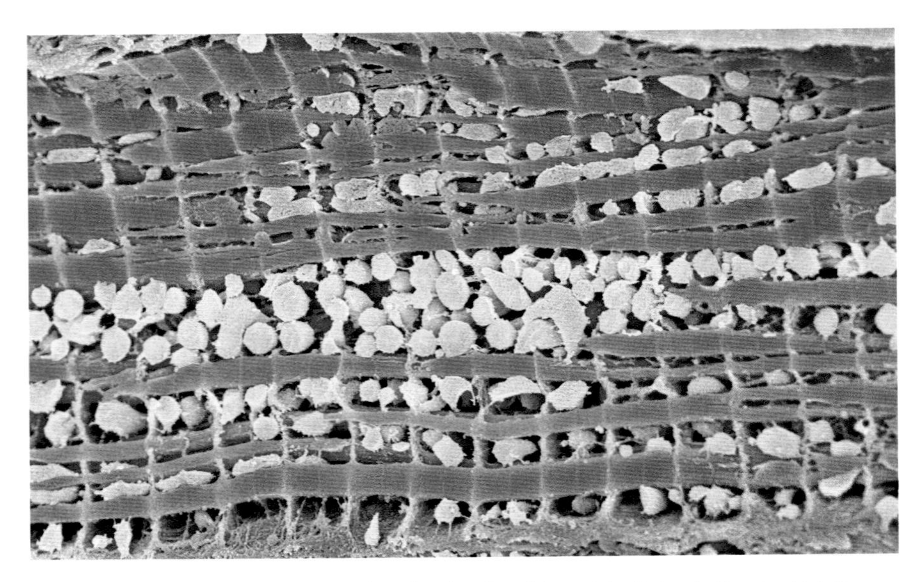

Inside a heart myocyte: stacked up within the cell are bundles of the proteins that make it contract (pink), while wedged in-between are the miniature power stations of the cell, the mitochondria (yellow).

Things are a bit more complicated than that, because the ventricles need to contract from the bottom up, so the electrical wave is quickly carried down from the atria to the bottom of the heart and then released. The heart valves are supported by fibrous rings that also act as electrical insulators between the atria and ventricles – stopping the electrical wave getting through – except for one small area, where another group of specialized cells leads into a bundle of cells that reach down to the bottom of the heart. This bundle branches into fibres which end just under the inner lining of the heart, where the cells release their electrical charge, causing the ventricles to contract from the bottom, squeezing blood upwards into the large arteries that take blood away from the heart (the aorta from the left side and the pulmonary trunk from the right).

The electrical activity of the heart can be picked up using electrodes on the outside of the chest – this is how an electrocardiogram (ECG) works. The line or trace of the ECG shows the electrical wave passing across the atria, then the ventricles, then the resetting of the heart's charge before the cycle happens again. Alterations to the ECG trace can show physicians where there might be damage to the heart: for example,

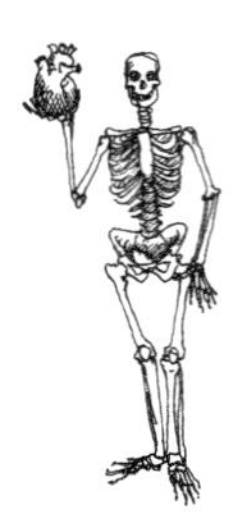

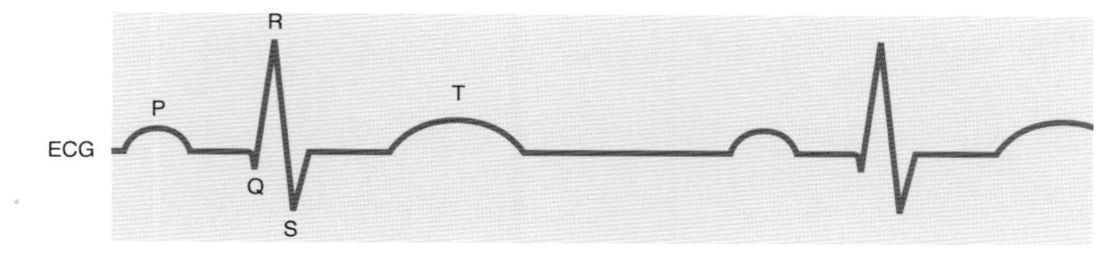

An ECG trace: the 'p wave' is the electricity passing across the atria . . . then a pause . . . then a wave of electricity across the ventricles ('qrs'), and the 't wave' appears as the heart is reset.

part of the trace may be changed when an area of heart muscle has died in a heart attack.

Sometimes the wiring of the heart can be faulty. This might be something you're born with or the result of damage, as in a heart attack. This can lead to abnormal rhythms (arrhythmias), which can cause dizziness, palpitations and fainting. Some arrhythmias can be very serious indeed: if the carefully coordinated sequence of contraction of the heart's chambers is disrupted, the heart may not pump hard enough (heart failure) or may indeed fail to pump at all.

The heart valves keep the blood flowing in the right direction. When the heart is relaxed between beats, blood pours into the atria, and down through the tricuspid and bicuspid valves, into the ventricles. When the atria contract, more blood is actively pumped into the ventricles. Then the ventricles contract and, as the pressure builds, the tricuspid and bicuspid valves close to stop blood flowing backwards into the atria, forcing the blood to leave the heart through the aorta and the pulmonary trunk artery. When the ventricles relax again, blood tries to fall back down into the heart but is stopped because there are also valves at the start of the arteries. These valves, the aortic and pulmonary valves, both have three flaps, which fill like pockets as the blood falls back, pressing their edges together and closing the valve.

When you listen to the heart beating, it's not the muscular contraction of the heart that you hear: it's the snapping shut of the valves. The tricuspid and bicuspid valves snap shut at the same time, making the first heart sound 'lub', then the aortic and pulmonary valves also close together 'dub'. So the heart sounds like this: 'lub-dub . . . lub-dub . . . lub-dub'.

The seat of the emotions

The heart used to be thought of as the 'seat of the emotions' and you can understand why. Although the heart controls itself, signals from the brain (but not the conscious bit) can speed it up or slow it down. Signals sent along the autonomic nerves cause the heart to speed up when you're stressed or scared, or to drop to a slower beat when you're feeling calm.

The heart slows down or speeds up its pumping to meet the needs of the tissues. If you're frightened, then your body prepares for flight or fight. Your muscles will be needed for either of these reactions, so autonomic nerves – and the hormone adrenaline – prime the heart to speed up and pump more blood to the muscles in readiness. It may feel like an 'emotional' reaction, as your heart thumps in your chest, but it's a very logical mechanism, to give you the best chance of survival. The autonomic nerves sometimes produce a strange but normal heart rhythm in children and young adults, where it speeds up as you breathe in, and slows as you breathe out.

As well as speeding up when you're stressed or scared, the heart rate also increases to get more oxygen to tissues during exercise – including itself. As the heart is a muscle, it also responds to regular exercise like any other muscle in your body by getting bigger and stronger. Athletes tend to have a slow heart beat – sometimes fewer than 60 beats per minute – because their training makes their heart more efficient.

How oxygen gets to your heart

To keep all this activity going, the heart needs a supply of richly oxygenated blood. This comes through the two coronary arteries, which spring off from the root of the aorta, just above the aortic valve. The left and right coronary arteries encircle the heart and send branches right down to the bottom of the heart. As long as the heart is pumping, all the cells in all the tissues of your body will have a constant supply of vital substances. But if it were to stop, everything is in trouble. The brain, the most oxygen and energy-hungry organ in the body, can't last more than a couple of minutes without its supplies. Unless someone quickly gets the pump working again, the brain will shut down, permanently.

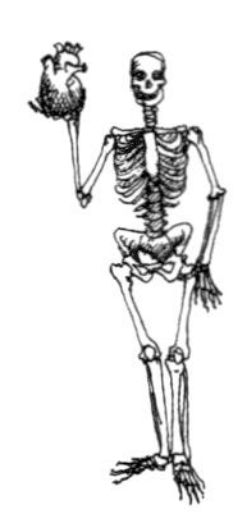

Common heart problems, and how to prevent them

The most common condition that affects the heart – coronary heart disease, the disease that produces a heart attack – is heavily influenced by lifestyle and diet. Coronary heart disease is the biggest health problem in western society and although some people are genetically predisposed to it, we can all lower our risk by living more healthy lives and eating a healthier diet. The risk of other common cardiovascular conditions, such as high blood pressure, arrhythmias, heart failure and deep venous thrombosis (DVT), can also be reduced by healthy living.

Coronary heart disease

Knowing a bit about the anatomy of the heart is key to understanding what happens in a heart attack, and there's a bit of a design flaw here. As I described earlier, the two coronary arteries branch off to supply richly oxygenated blood to all parts of the heart – but they don't really connect up with each other. So if a coronary artery, or one of its branches, gets blocked by a clot, there's not any other way of getting blood to the bit of heart muscle beyond the block: it gets starved of oxygen and dies. The technical term for a heart attack is 'myocardial infarction' (often shortened to MI). The myocardial bit means 'heart muscle' and infarction – a great word – literally means 'stuffed in' and refers to the clot getting wedged into the coronary artery.

A narrowed artery can cause problems, even without being completely blocked by a clot, because the amount of blood that can flow through it is restricted. When the heart is pumping hard, for example during exercise, it needs more blood than when you're just sitting down. If the arteries are narrowed, the heart muscle may not get all the blood (and therefore oxygen) it needs – and it gets cramp. This cramp is felt as a shooting pain, which often seems to be coming from the shoulder or neck (because the same nerves supply the shoulder and neck as well as the heart, and the brain gets confused about where the pain signal is

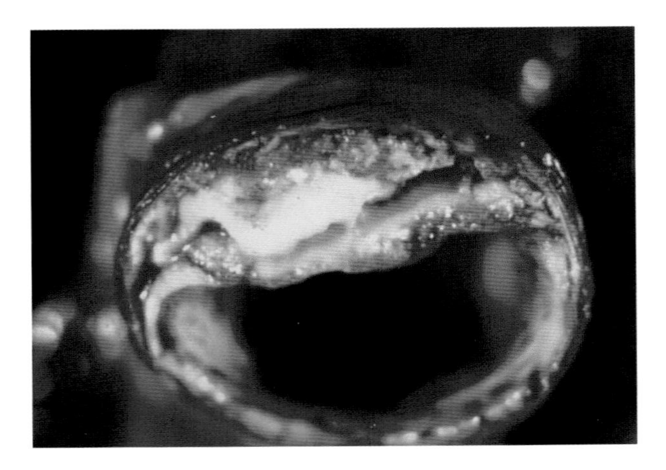

Atherosclerosis: an artery narrowed by a yellow, fatty deposit of atheroma in its wall.

coming from). This heart cramp is angina (Greek for 'strangling'). The narrowed coronary arteries mean that as the heart tries to work harder, it can't supply itself with enough oxygen, and is literally strangling itself.

So how can you look after your heart? A good place to start would be to avoid coronary heart disease. We need first, to keep our coronary arteries nice and wide, and second, to reduce the likelihood of clots forming in the blood.

You're much more likely to get a clot completely blocking an artery if that artery has already been narrowed by fatty plaques being laid down in its walls. The stuff of these fatty plaques is known as 'atheroma' (Greek for 'porridge') and isn't on the inner surface of the artery but actually within the inner lining (the 'tunica intima': a wonderful phrase which basically means the 'underwear' of the artery). Atherosclerosis (the narrowing of the arteries due to atheroma) seems to begin with damage to the lining of the artery, which could be due to infection, toxins like cigarette smoke, high levels of glucose or mechanical damage from high blood pressure. White blood cells home in on the damaged lining and get inside it. Once inside, they undergo a strange transformation, developing an odd taste for cholesterol, absorbing large amounts until they become great, puffed up bags, full of cholesterol. In this condition, they are called 'foam cells', because they look like swollen foamy bubbles under the microscope.

Massed together, the foam cells form a fatty streak on the inside of the artery, which eventually grows to become a fibrous plaque. The narrowing of the artery by the plaque isn't dangerous by itself – the real danger looms when the plaque becomes unstable. If the surface of the plaque cracks, a thrombus (Greek for 'clot') will form over the wound (rather like a scab forming when you cut your finger). This thrombus

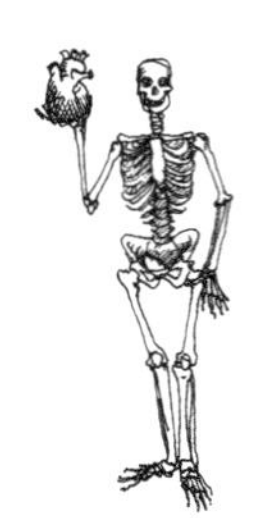

can form very quickly and may completely fill and block a coronary artery, causing a heart attack. Surges of blood pressure during heavy exertion or stress may cause a plaque to crack and lead to a clot forming. Another way for an artery to get blocked is by bits of thrombus that have formed elsewhere arriving in them. A bit of thrombus that breaks off, as a mobile clot, is called an embolus (Greek for a 'plug'). If an embolus is carried into a vessel already narrowed by atheroma, it may get wedged in, blocking the artery.

Good cholesterol . . . bad cholesterol: If atherosclerotic plaques contain cholesterol, cutting down on cholesterol in our diets – for instance, by eating fewer eggs – should stop us getting heart disease. This was the kernel of the diet-heart hypothesis that emerged in the first half of the twentieth century. Cholesterol quickly became Public Enemy Number One. But then the picture got more complicated, as biochemical studies revealed that cholesterol travelled in the blood in different packages: as light or heavy globules. It seems that the balance of these two types, as well as total blood cholesterol, is important in heart disease. Relatively high levels of 'bad' cholesterol (in light globules or low density lipoprotein) increases the risk of heart disease, whereas relatively high levels of 'good' cholesterol (in heavy globules or high density lipoprotein) reduces it.

The amount of cholesterol you eat seems to have little effect on the total cholesterol in your bloodstream or indeed how it is packaged. What appears to be more important is the amount of fat and, in particular, the amount of saturated fat, in your diet. Saturated fats tend to be solid, and are found in butter, lard, fatty meat, cheese, cakes and chocolate; unsaturated fats tend to be liquid – olive oil and rapeseed oil are high in unsaturated fats. Levels of bad cholesterol are increased by saturated fatty acids (produced from digested saturated fats) and especially by trans-fatty acids. ('Trans' describes a very detailed bit of the molecular architecture: it means there's a double bond between two carbon atoms in a chain, with hydrogen atoms stuck on opposite sides of the carbons.)

All you really need to know about trans-fatty acids is that they have an adverse effect on the balance of cholesterol globules, and that you should therefore try to avoid them. Many of the trans-fatty acids in our diets are artificial: formed by deep frying, by baking and when vegetable oils are treated with hydrogen to make them more stable, such as in margarine. Trans-fatty acids also occur naturally in foods like milk, cheese, eggs and meat. The current UK recommendation is to eat less than five grams per day; you should be able to do this by concentrating on cutting down on fried and manufactured foods. The less the food you buy has been mucked around with, the better.

A Mediterranean diet: Cutting down on fat, and making sure you're eating the right sort of fats, can help to keep your heart healthy. While saturated fats are definitely linked to heart disease, switching to a 'Mediterranean diet' with olive oil and plenty of fruit and vegetables has been shown to reduce heart disease significantly. Fruit and vegetables are extremely important: they contain vitamins and antioxidants which reduce the risk of coronary heart disease, as well as many other diseases. Bad cholesterol is more likely to be absorbed into the lining of arteries if it has been oxidized; antioxidants prevent the cholesterol from being oxidized, so they slow down atherosclerosis. The UK Department of Health recommends eating five portions of fruit and vegetables per day: this mantra is a good rule of thumb but to some extent you can forget the counting and just eat as many fresh fruit and vegetables as you can!

Studies also suggest that specific foods in the Mediterranean diet, for example, garlic and onion, may have particular protective effects against heart disease, high cholesterol levels and high blood pressure. Nuts, in particular almonds and walnuts, are a very good source of antioxidants and help to raise the level of good versus bad cholesterol. Apples, onions and red wine (and tea!) contain flavonols, strong antioxidants that appear to significantly reduce the risk of death from coronary heart disease. It's interesting that, although the beneficial effects of eating fruit and vegetables are well documented, there doesn't seem to be any

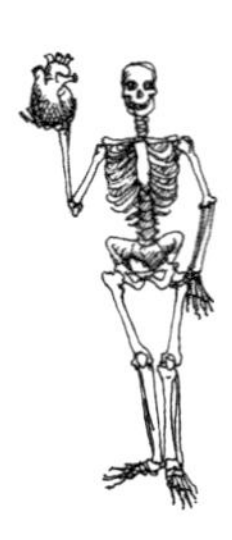

evidence that vitamin or antioxidant supplements have any effect in preventing or treating coronary heart disease. Pills won't make up for an unhealthy diet.

The traditional Mediterranean diet is also rich in fish. Fish oil is full of omega-3 fatty acids. The name 'omega-3 fatty acid' may trip lightly off the tongue of an organic chemist but what on earth is it? This is another bit of molecular architecture to get your head round – and believe me, I'm delving back into the mists of time, to my A-level chemistry and medical undergraduate biochemistry.

Well, fatty acids are essentially long chains of carbon atoms with hydrogen atoms attached. In an omega-3 fatty acid, three carbon atoms before the end (called 'omega' after the last letter of the Greek alphabet), there's a double bond between the carbon atoms. Omega-3 and omega-6 fatty acids are called 'essential fatty acids' because our own cells can't manufacture them: we have to get them in our diets. Essential fatty acids seem to have several positive effects in the body: they cause a relaxation and widening of arteries, helping to keep blood pressure low; they help keep glucose levels down, they reduce the tendency of blood to clot and they improve cardiac function. This adds up to a positive effect on the health of your heart: eating fish once a week may halve your risk of heart disease. There's no need to panic if you absolutely hate fish or are vegetarian. You can get the essential omega-3 fatty acids from plant sources too, such as flaxseed and rapeseed oil, green leafy vegetables and walnuts.

Alcohol has its place in the heart-protecting Mediterranean diet: it raises levels of good cholesterol and lowers bad, so reduces the build-up of atheroma in arteries. It also has an anti-clotting effect. However, the key is 'all things in moderation' – having one or two drinks per day has been shown to lower the risk of heart disease by up to 40 per cent but drinking your entire quota for the week in one go is bad for the heart (and just about every other organ). As Phil Hammond put it, in his funny and informative book *Trust me, I'm a Doctor*: 'Drinking heavily on Saturday will not build up your resistance for the rest of the week'. From the heart's point of view, there is no benefit in having days off drinking, as the positive effects on the heart and blood last for only a day. Which is not to say teetotallers should take to the bottle: for people who don't drink alcohol, the cardiovascular benefits of drinking aren't spectacular enough to suggest that they should go back, especially if they've stopped drinking because of dependency or liver problems.

The heart-friendly effect of a moderate alcohol intake might explain the so-called 'French Paradox'. In spite of eating lots of cholesterol and fat, French people who drink wine daily have a lower rate of heart disease than people in other western countries. It's no small difference: three times as many people die from coronary heart disease in Glasgow than in Toulouse. It's not just France that enjoys this blessed relief: the hearts of the whole of southern Europe are healthier than ours, up north. The effect has been traced to red wine in particular; research shows that the protective effects of wine on the heart outweigh those of beer or spirits. As well as containing alcohol, red wine is also a good source of antioxidant flavonols, which put a spanner in the works of the first stage of atherosclerosis.

Exercising your heart: Exercise is extremely important for keeping your heart healthy. Regular aerobic exercise (such as walking, swimming or cycling) increases the fitness of the heart muscle and helps reduce blood pressure. It also has other heart-friendly effects; like a moderate alcohol intake, regular exercise lowers the total amount of fat and cholesterol in your blood, while raising the level of good cholesterol. It also makes your blood less liable to clot. A recent study of apparently healthy young men and women showed up the dramatic differences between those who were physically active and those who led a sedentary, sofa-centred life. The people in the study were all under 40 years old but the arteries were already harder in the sofa-bound group. The risk of coronary heart disease is about twice as high in inactive people compared with that of active people.

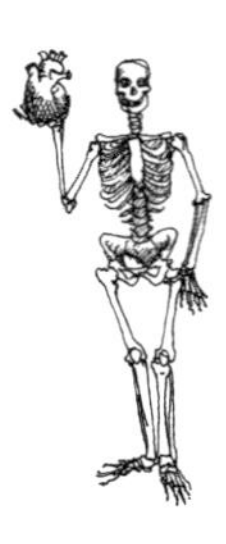

Exercise is also important for people who already have coronary heart disease: exercise can reduce the risk of death by 25 per cent – but you should

Cycling is just one way of keeping your heart trim (but I'm not sure how this chap is going to manage cycling uphill, with no gears).

talk to your doctor about how much exercise you should be doing. Regular exercise lowers your risk of having a heart attack in the first place and it also reduces your risk of dying if you have a heart attack.

However, heavy exertion can also trigger cracks in plaques, leading to clot formation and heart attack. How can this risk be weighed against the benefits of regular exercise? The real risk occurs when people who are usually sedentary exert themselves unusually heavily. Not only is a sudden rush of blood likely to disrupt any plaques, but clots form in these people more easily, and their bodies are not very good at naturally dissolving clots. In people who are healthy and active on a daily basis, the risk of having a heart attack when they exert themselves is lower. Regular exercise puts the brakes on the speeding-up of the heart and reduces the surge in blood pressure. It also means the blood is less liable to clot and natural clot-busting mechanisms are stronger. Any heavy exertion – including sex – can be risky if you are very unfit, but you can lower your risk of having a heart attack due to sexual – or any other – exertion by keeping generally fit and active. There's obviously a link between lack of exercise and obesity, so it's not surprising that people who are overweight have a higher risk of heart disease.

Stress: All the factors I've mentioned so far have been physical, but mental stress is a major risk factor for angina and heart attacks. Mental stress can exert such a large effect that it can actually cause a restriction of blood supply to the heart muscle when someone is physically at rest. Mental stress sparks the sympathetic nervous system into making the heart beat stronger and faster, but given that you don't actually need to run anywhere, this isn't

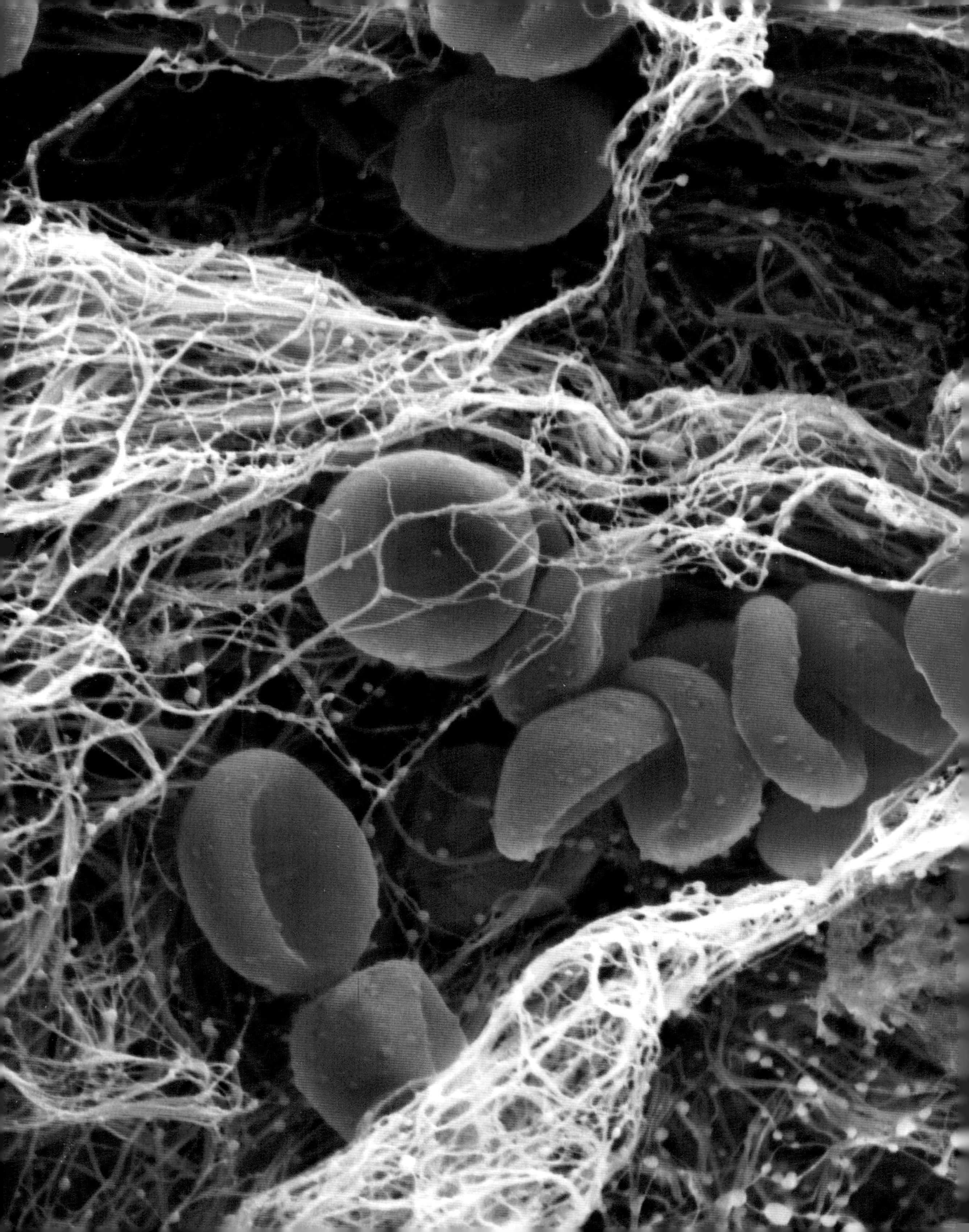

a useful response. It can also cause abnormal heart rhythms, which can result in a lack of blood getting to the heart muscle – even if the arteries aren't physically blocked – because the heart isn't pumping effectively. Mental stress also increases the tendency of the blood to clot and decreases the clot-busting ability of the blood, meaning that clots are more likely. Episodes of anger, fear and anxiety increase the risk of arrhythmias, angina and heart attacks.

Smoking and drugs: Using cocaine or marijuana increases the risk of heart attacks. They both cause the heart to speed up and cocaine also speeds up atherosclerosis and narrowing of the arteries. Cigarette smoking is an important – if not the most important – factor in coronary heart disease. It's just plain bad news for your heart and arteries. It causes inflammation of the lining of arteries, stopping them functioning properly. It has the opposite effect to red wine, oxidizing bad cholesterol and making it more likely to be absorbed into the walls of arteries. And it makes your blood clot more easily. The risk of heart disease increases in proportion to how many cigarettes you smoke each day and how deeply you inhale the smoke. Smoking is the biggest single cause of death and illness in the UK – and it's completely avoidable. If you need a financial incentive, try visiting the NHS Direct website, look up smoking and click on 'Find out how much you'll save by kicking the habit'. It's also important to remember that passive smoking is almost as harmful as active smoking in increasing your risk of heart disease.

It's down to you: Although your genetic make-up plays a role in disease, you can have a very large impact on your risk of developing many diseases. Eighty per cent of the risk of coronary heart disease is down to five factors: high levels of bad cholesterol, smoking, diabetes, high blood pressure and midriff obesity. These are all 'potentially modifiable' factors; in other words, unlike your age or your sex, which you can't change, you do have control over these factors. You can reduce the risk that all these factors bring to the mix if you eat a healthy diet with plenty of fruit and vegetables, exercise regularly, give up smoking, and drink alcohol in moderation. The evidence

Clotted blood: the protein fibrin has formed a mesh, along with platelets, trapping doughnut-shaped red blood cells. In certain conditions, such a tangled net could quickly build up to block a coronary artery.

shows that the combined effect of a healthy diet and lifestyle is larger than the effects of each independently. A healthy living 'package' is the key to a healthy heart. This applies just as much if you already have coronary heart disease. There is overwhelming evidence to show that changes to lifestyle and diet, especially stopping smoking, can stop people who have heart disease from getting worse and can significantly reduce their risk of dying from a heart attack.

High blood pressure

Having high blood pressure also increases the risk of heart disease. Surges of blood pressure can disrupt atherosclerotic plaques, and high blood pressure can have lasting effects on the heart muscle. The heart has to pump harder, and the left ventricles of people with high blood pressure are often enlarged and the extra heart muscle demands more oxygen.

High blood pressure (or hypertension) begins with problems in the blood vessels, with their lining and with the muscle in their walls. The lining of the arteries can be damaged by free radicals, usually overly reactive oxygen molecules. Free radicals can also deactivate nitric oxide, a useful little molecule that helps to keep your arteries open. Without nitric oxide, the arteries constrict, and because you're trying to push the same amount of blood through them, the pressure inside them increases. Free radicals produce other effects that also contribute to high blood pressure: they damage the lining of arteries, oxidize bad cholesterol so that it's more readily taken up into the lining, and lead to higher glucose levels in the blood. Antioxidants combat the effect of free radicals by neutralizing these molecular time bombs, letting more nitric oxide stick around and so keeping your arteries wider and the blood inside them at a lower pressure.

High blood pressure increases your risk of angina and heart attacks and lots of other diseases. When the doctor measures the blood pressure in your arm, and finds that it's high, this means it's high in all your arteries and in all your organs. High blood pressure means you are six times more likely to have a stroke and more likely to have narrowed arteries in your limbs; it also damages your kidneys.

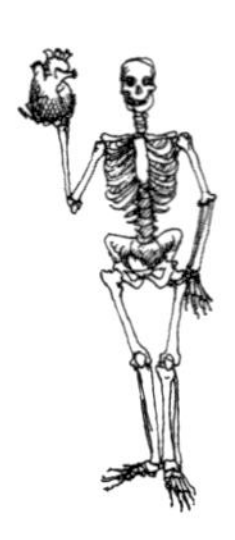

Keeping your blood pressure down is important. The lifestyle changes that reduce your risk of developing high blood pressure are largely similar to those for avoiding coronary heart disease: it's important to eat a healthy diet, get regular exercise, lose excess weight and stop smoking.

Reducing the amount of salt in your diet and eating more fruit and vegetables, nuts and dairy products low in saturated fat, can bring down high blood pressure. Cutting down on caffeine and alcohol (there's evidence that women who drink more than three units per day and men who knock back more than four are likely to have raised blood pressure) can also help. There are drugs (such as diuretics and ACE inhibitors) that are very effective in lowering high blood pressure but the result will always be better if you also make healthy changes to your life and diet. In some cases, dietary changes alone can be as influential as drug therapy.

The key to keeping your blood pressure down seems to be to give your body something a bit more like the diet it evolved to live with thousands of years ago: high in potassium, fruit and vegetables, lean meat and fish (omega-3 fatty acids also stimulate nitric oxide production, helping to keep your arteries wide), but low in sodium, saturated fat and cholesterol. Vitamin C and plant flavonoids (found in numerous fruit and vegetables, red wine, soy and liquorice) are potent antioxidants, which are very good at scavenging for free radicals – and relaxing the arteries. Don't think you have to rush out and buy supplements – there are plenty of these antioxidants in whole foods (mainly fruit and vegetables); eat plenty of those and you should get all the vitamins and antioxidants you need. Pound for pound, blueberries are more potent an antioxidant than neat vitamin C!

There are some foods that seem to be particularly effective at lowering blood pressure: celery, hawthorn berries (bizarrely), garlic, seaweed, tuna, sardines, egg yolks and fibre. The natural combinations of vitamins and antioxidants in whole foods are generally superior at preventing and treating high blood pressure to singled-out components in supplements. However, there may be a case for taking supplements if you have high blood pressure and are deficient in a particular vitamin or nutrient. For example, co-enzyme Q-10 (often shortened to Co-Q-10) is an important antioxidant, and low levels are often associated with high blood pressure, coronary heart disease and diabetes. It can be difficult to get it in your diet, so if your levels are deficient, your doctor may recommend that you take Co-Q-10 supplements.

The body needs, at a minimum, about half a gram of sodium per day. The average American eats ten times that and some people eat more than 20 grams per day – 40 times the minimum required amount. In people with high blood pressure, reducing the salt intake has been shown to be very effective in lowering it. But severe restriction of sodium can also lead to deficiencies of other substances like potassium and calcium, which may have the effect of raising blood pressure.

It's important to get the balance right; a gradual reduction in sodium intake, together with plenty of fruit, vegetables, low fat dairy products and fibre, has been shown to be most effective in reducing blood pressure. There's masses and masses of literature on the relationship between salt and high blood pressure but there's still no consensus on how it works. For healthy people, the salt spectre may be overplayed; salt is an important part of our diets. The sodium you get from salt is an essential component in cells and the fluid around the cells and it plays an important role in muscle contractions and nerve impulses. Salt seems to suffer from the same unfortunate labelling that cholesterol ended up with but these substances aren't bad for you unless you consume much too much. In sensible quantities, they are important components of a healthy diet.

Varicose veins

So far, I've concentrated on the heart and arteries. But now for a quick look at veins and common problems that affect them.

Varicose is simply Latin for 'twisted', and varicose veins develop when the veins just under the skin, particularly those of the leg, become swollen and tortuous. Veins have little valves inside them; flaps of tissue that stop blood flowing the wrong way. The blood in arteries, flowing away from the heart, has a strong current, and waves or pulses of pressure surge through it each time the heart beats. The blood in veins, returning to the heart, has a much more sluggish current. In the leg veins, the blood has to flow upwards

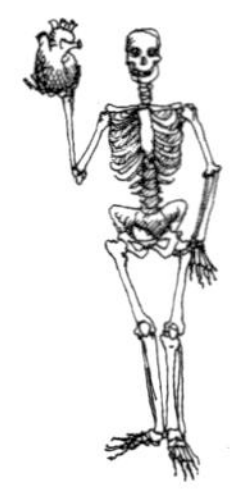

This coloured angiogram (where a dye has been injected into the veins before an X-ray was taken) shows twisting and enlarged varicose veins running up the back of the calf.

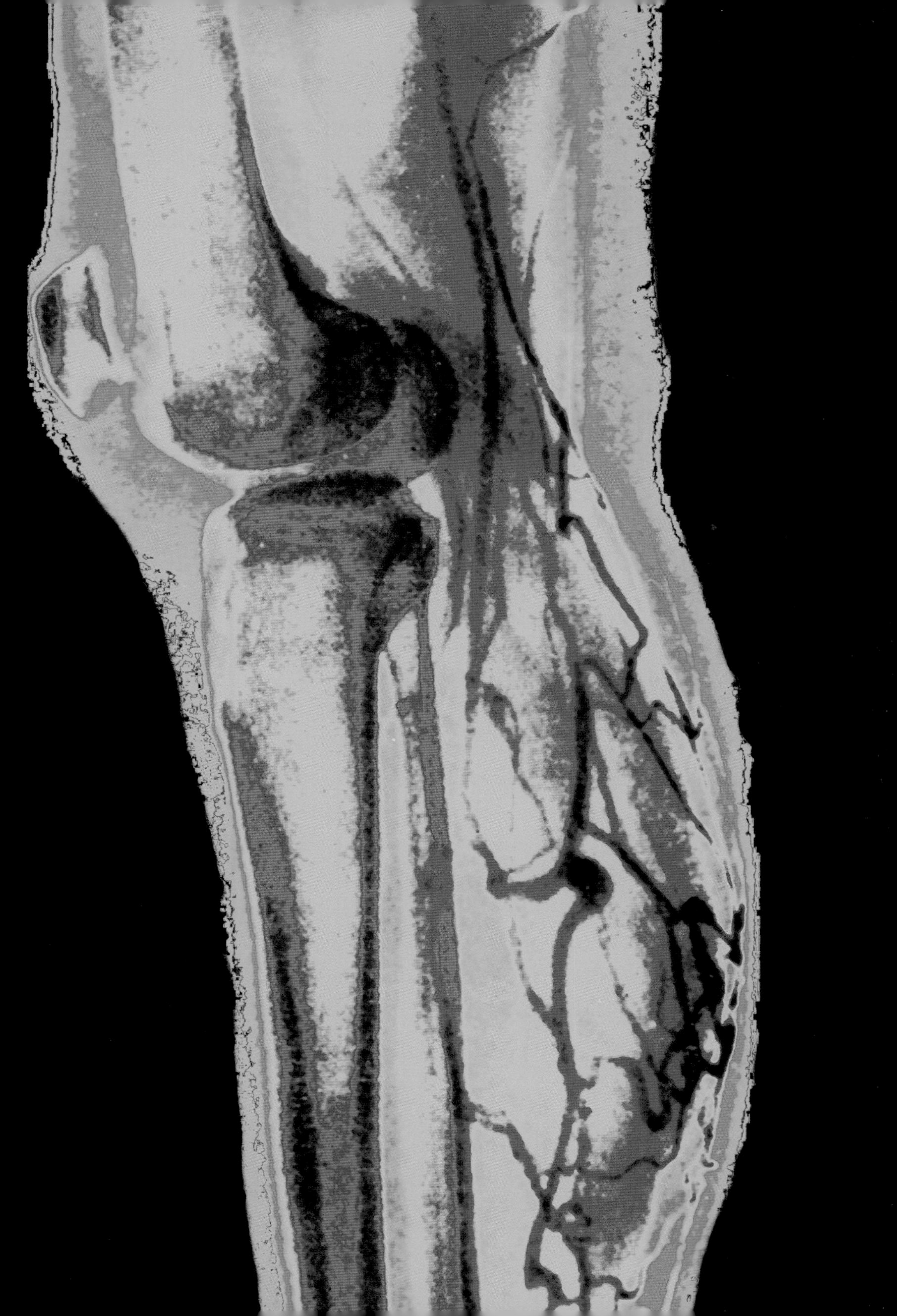

(against gravity) and the current sometimes slows to a standstill. The deeper veins of the leg run between muscles; when those muscles contract, it helps squeeze along the blood inside the veins. Valves inside the veins make sure that the blood is squeezed in the right direction: up towards the heart, instead of down towards the foot.

There are connections between the superficial veins just under your skin and the deeper veins between the muscles. There are also valves here, to make sure that blood flows from the superficial to the deep veins – and not the other way around. But those valves can become faulty – meaning that blood then pools in the superficial veins, stretching them and stretching their valves, so that more blood pools. This is how varicose veins form. They are more likely to form if there's something pressing on veins higher up and making it harder for the blood to run up the leg. This might happen if you get too fat; it also becomes a problem during pregnancy. To reduce your risk of having varicose veins, keep your weight down and stay active – to keep that muscle pump in your legs working.

Deep venous thrombosis

Deep venous thrombosis (DVT) pops up a lot in the media because of its link with long-distance flying. DVT happens when a clot forms in one of the deep leg veins, and it is *extremely* painful. A particular worry is the risk that bits of the clot might break off and if they do, they could go up the veins to the heart, through the right atrium and ventricle and off into the lungs, where they can get stuck in one of the smaller arteries. This is called a 'pulmonary embolism'. Some of the risk factors for thrombosis are inherited, such as a tendency for your blood to clot, but others are things you can do something about, including smoking, obesity and a lot of sitting about. Long air flights may be a particular risk, as you end up sitting still for long periods of time.

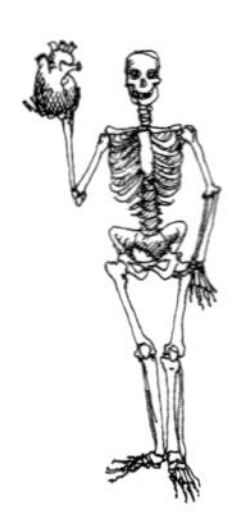

Five ways to keep your heart healthy:

 Don't smoke.

Eat a Mediterranean diet, with plenty of fruit and vegetables, wholegrain cereals and bread, low-fat dairy products, oily fish, lean meat, and nuts (especially almonds and walnuts). Cut down on fat and replace saturated fat with unsaturated oils (swap lard for olive oil!). And cut down on salt.

Drink alcohol – in moderation (one to two units per day for women and one to four for men) – but don't binge drink! There are also lots of antioxidants in cranberry and purple grape juice.

Get plenty of physical exercise – aim for at least 30 minutes of aerobic exercise per day (for example, walking, cycling, swimming or gardening).

 Try to avoid stress.

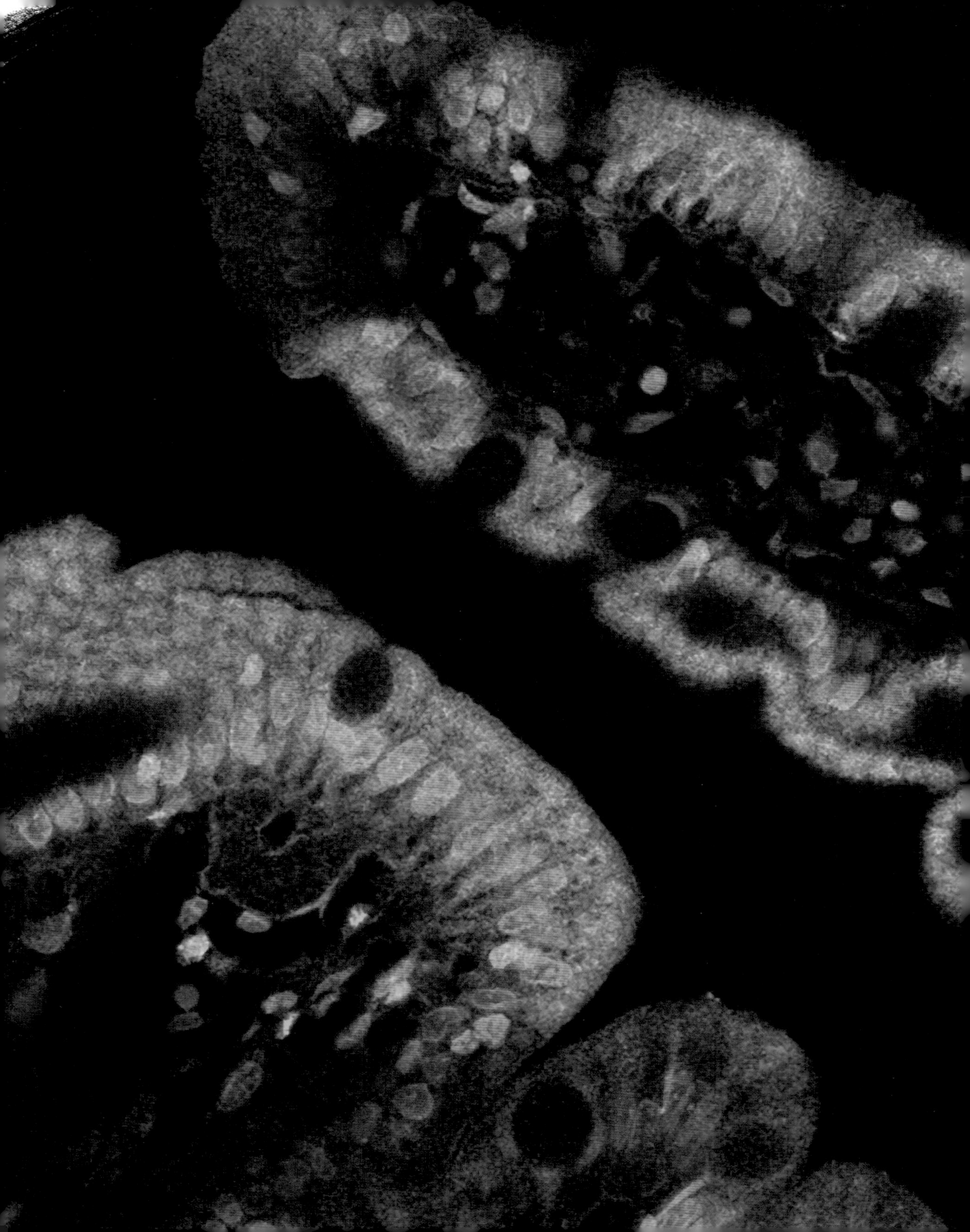

THREE

THE STOMACH AND INTESTINES

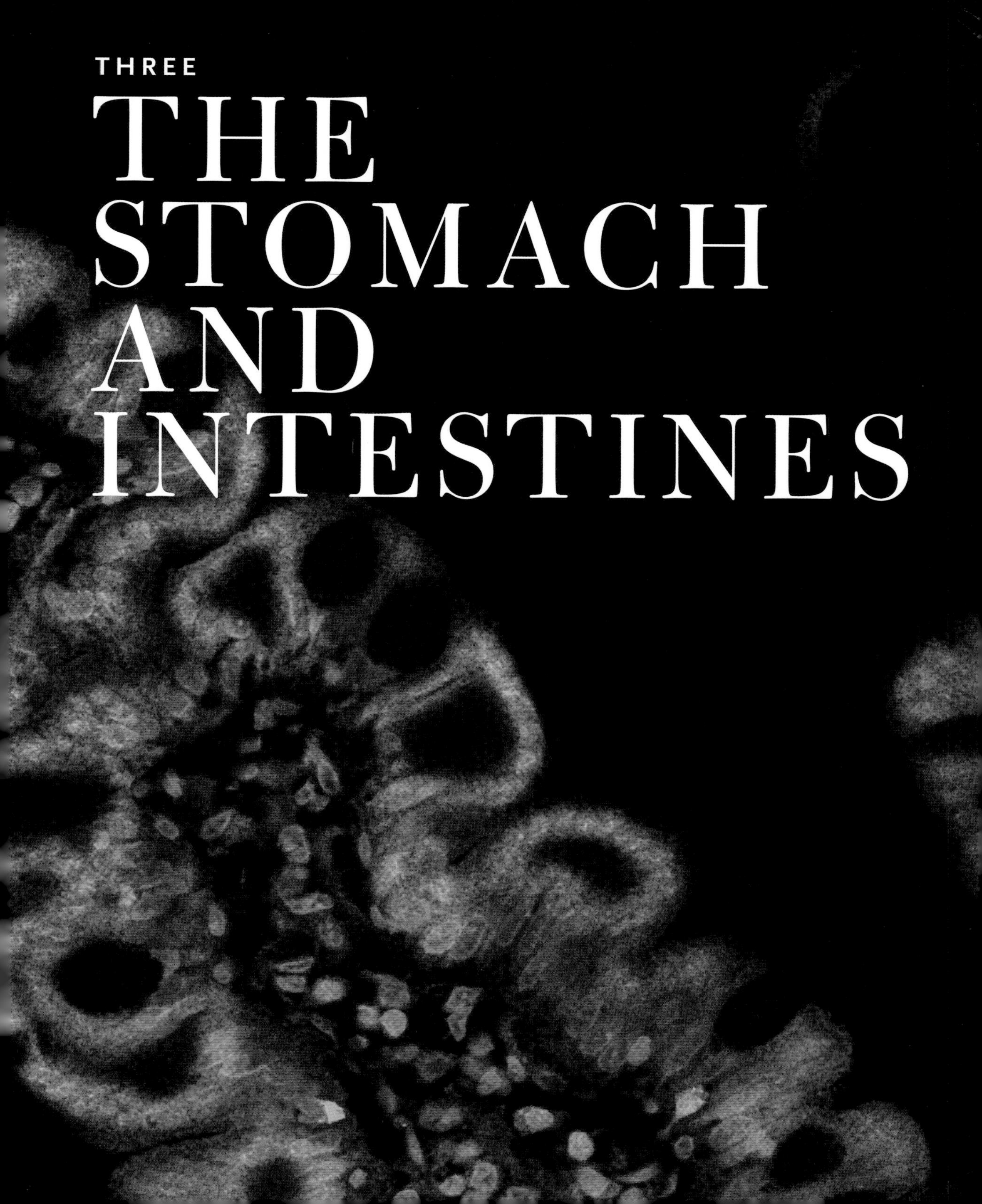

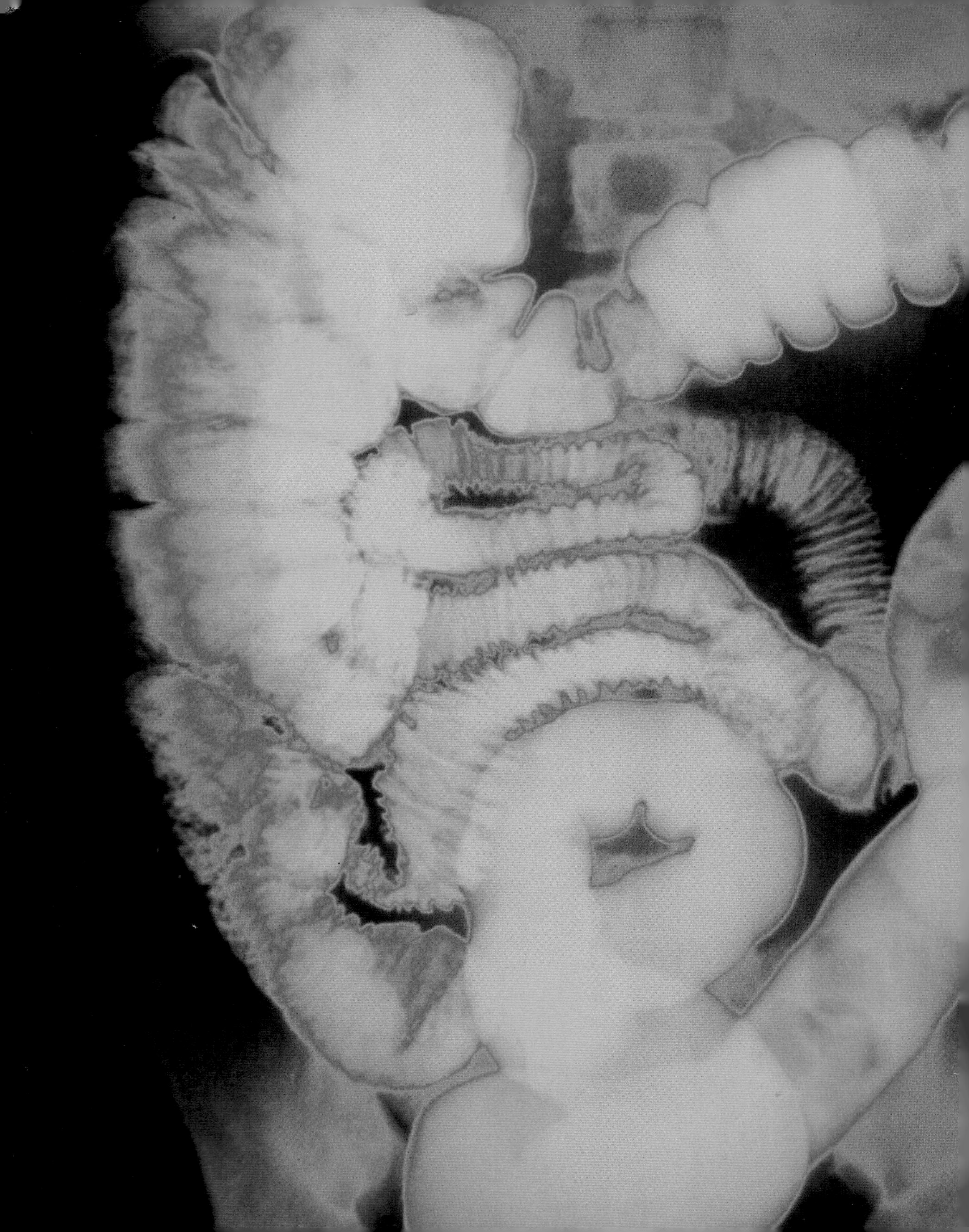

The stomach and intestines are just part of the long tube that stretches all the way from your mouth to your anus, but they are definitely the 'business district', where most of the digestion and absorption of food goes on.

Getting digested along the way, it usually takes one to two days for ingested food to travel all the way from the mouth to the anus, depending on how much dietary fibre (plant material) there is in your diet; transit time tends to be quicker in vegetarians.

Together, the stomach and intestines form the gastrointestinal tract, which takes chunks of stuff and breaks them down into smaller bits that can be absorbed into the bloodstream and used as fuel or building material by the body. This is quite a strange concept – animals, like us, take parts of the outside world, internalize them, and turn them into parts of our body. There's something quite parasitic about the way we feed on our surroundings.

Previous pages **A villus in the duodenum, stained so that the cells' nuclei glow orange.**

Left **Barium in the bowels: barium makes the inside of the intestine show up on an X-ray – the large tube around the periphery is the colon; the zigzagging coils in the centre are the small intestine.**

Where are the stomach and intestines?

The stomach sits high up in the abdomen, mostly tucked under the ribcage on the left. The oesophagus carries food from the throat, down through the thorax and through the diaphragm, emptying it into the stomach. The stomach leads on into the intestines, a long tube which is neatly coiled up into the middle and lower parts of the abdomen. The lowest coils rest in the pelvis, sitting on top of organs such as the bladder and uterus.

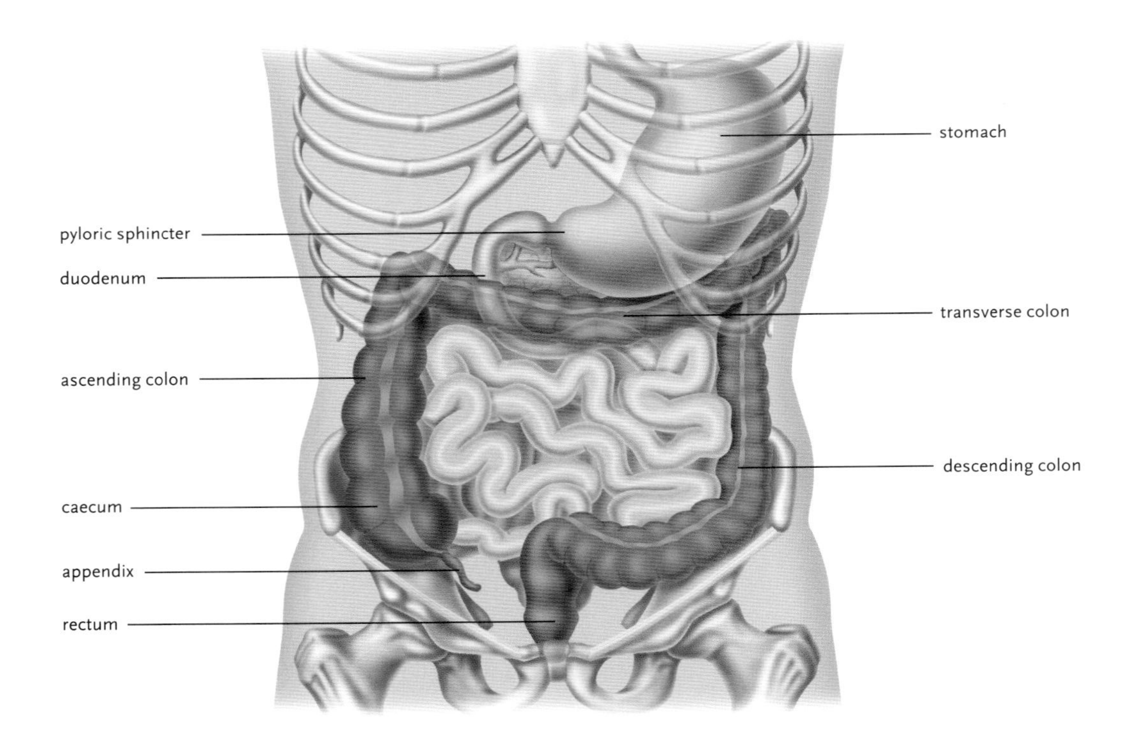

Omnivore guts: humans have a relatively simple gut; not for us the four-chambered stomach of the cow or the impossibly huge caecum of the horse.

How the stomach and intestines work

The stomach is a muscular bag that can expand enough to hold a whole meal. It churns up the food, starts to digest it and releases it, a little at a time, into the long, narrow tube of the small intestine – which is about six metres long. As the food passes along this tremendous length it gets digested down even more and the products of digestion are absorbed into the blood. (This blood then goes to the liver, which decides what to do with the nutrients: which to use now, which to store, and so on.) After the small intestine, the squelchy digested food goes into the large intestine, which is wider than but not as long as the small intestine. Here, precious water gets sucked back into the bloodstream, leaving solid stuff behind. This goes into the rectum, which can expand to hold quite a bit before the need to empty your bowels becomes a priority.

The British have a reputation for being obsessed with their bowels but this isn't an entirely misplaced concern. Cancers of the stomach, colon and rectum account for nearly 20 per cent of all cancers. Around the world, almost a million people get colon cancer every year and nearly half a million die from the condition. Not surprisingly, diet has a large part to play in many of the common medical conditions affecting the stomach and intestines.

How food travels through your gut

Getting from large lumps of food to molecules that are small enough to be absorbed into the bloodstream involves both a physical reduction in size and a chemical breakdown of the large, complex molecules into their smaller building blocks.

The physical digestion of food starts in the mouth, where the job of the teeth is to smash up chunks of food into small lumps that are mixed with saliva for swallowing. Saliva not only moistens the food and eases its passage on the first part of the journey down the oesophagus and into the stomach, but also contains enzymes (special proteins that speed up chemical reactions) which chemically digest the food, cracking down starches into sugar molecules.

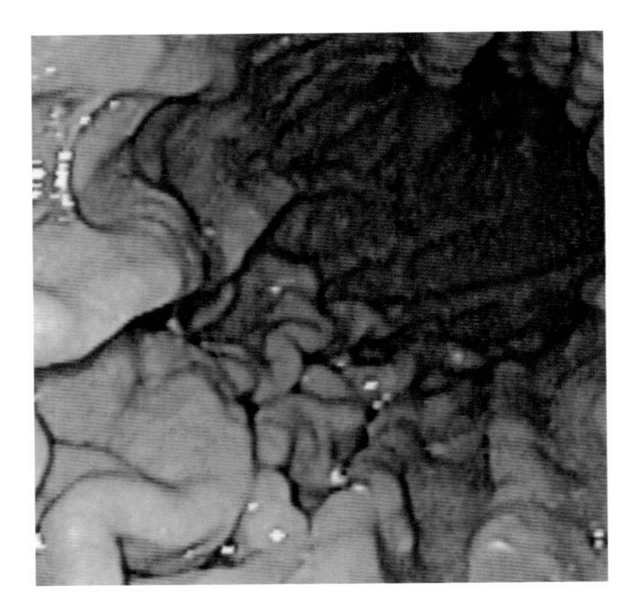

Wrinkled rugae: the folds of the stomach's lining allow it to expand when a meal comes its way.

The chewed-up food is swallowed into the stomach, a holding vat where your meal is detained before it goes on to the next stage of digestion in the intestines. But it doesn't just hang around: the stomach has its own role to play in digestion. In the stomach, your food is bathed in a concoction of hydrochloric acid (which helps to kill any bacteria that you happen to have ingested along with your food) and enzymes (which in this case digest proteins). The stomach protects itself from its own strong acid by lining itself with a thick layer of mucus.

The stomach is stretchable and expandable – open up an empty stomach and you would see that the lining is folded into 'rugae' (Latin for 'wrinkles'), folds that get smoothed out as the stomach fills. The rugae are like the folds of a furled umbrella – but on the inside. The stomach's thick muscular wall contracts, churning and breaking up its contents – continuing the work of physical digestion that the teeth started. The stomach also produces chemical signals, which warn the intestines that food is coming their way.

The stomach slowly releases its contents, a little at a time, into the first part of the small intestine. At the exit from the stomach, the muscle thickens to make a sphincter (Greek for 'to draw tight'). This region is called the pylorus (Greek for 'gatekeeper'), so the ring of muscle is called the pyloric sphincter. It's a bit of anatomy that I knew about long before I went to medical school: my dad was unlucky enough to be born with an unusually narrow pylorus (known as pyloric stenosis). He had to have an operation to open up the sphincter, which left a long scar down his belly, but the procedure saved his life.

Pyloric stenosis has a strong genetic component and affects boys more than girls: 35 years later, my brother was born with the same problem. Very little milk got through his pylorus into his intestine, so he wasn't hungry and

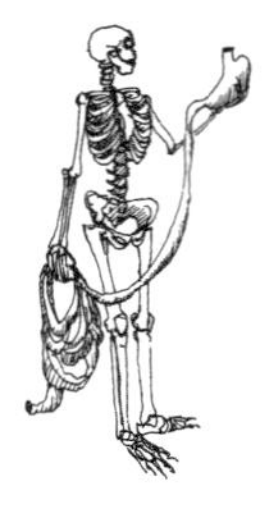

would vomit after every feed. He had to have an operation too, but surgery had progressed and he was left with a much smaller, neater scar.

When the pyloric sphincter opens, the semi-digested, liquid food squirts into the first part of the intestine: the duodenum. (Duodenum is the Latin for 'twelve'. It relates to an unconventional measurement: the duodenum is twelve finger-breadths long.) It is a curvy loop of intestine – it heads off to the right, turns down and then round to the left, then loops up and turns forwards into the next part of the intestine. Halfway down the duodenum, there's a nipple-like protrusion with a small opening. The pancreas and the gall-bladder empty their contents into the gut through this opening: pancreatic enzymes and bicarbonate to neutralize the stomach acid, and bile (produced in the liver and stored in the gall-bladder) to break down fat into droplets.

The food – by now quite liquidized – passes into a part of the intestine that is 'loose' inside the abdomen (though attached to the back wall by the membrane called a mesentery). The first couple of metres of this loose intestine are the jejunum (Latin for 'empty', perhaps because food passes through this bit quickly compared with the holding vat of the stomach) and

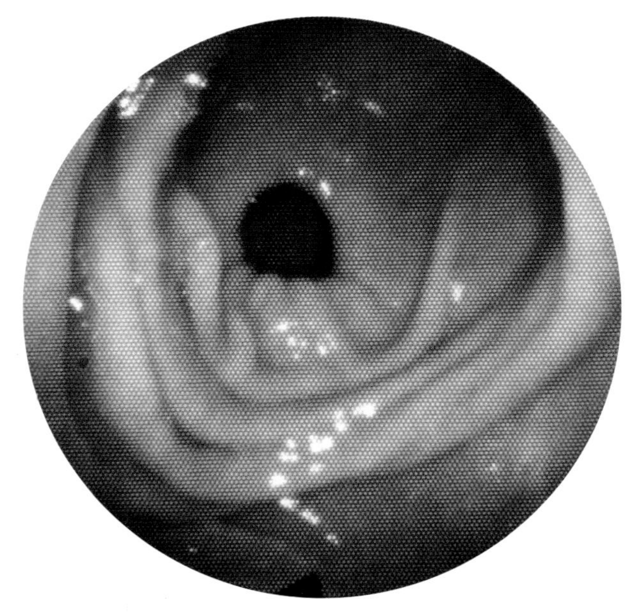

The gatekeeper: the exit from the stomach narrows down to a small hole that opens to allow small amounts of digested food through into the duodenum on the other side.

the rest is the ileum (Latin for 'entrails' and Greek for 'rolls'). This long length of intestine rests in coils in the lower part of the abdomen. It's a muscular tube, with an outer layer of muscle fibres arranged along its length and an inner layer arranged in a circular fashion. These two muscle layers, contracting alternately, push the liquid food along. It's quite amazing to see this happening: when a patient's abdomen is opened up in an operation, the contractions passing along the coils of gut look just like the movement of an earthworm. The word for this special type of muscle contraction, peristalsis, which describes a constriction moving along something, is used both for the movement of the intestines and the movement of worms.

Like other organs of the body whose function we don't have to think about consciously, the gut gets on with its job of moving food along, digesting and absorbing it. It has its own set of nerves (sometimes called the 'gut brain') embedded under its lining and in its muscular wall that keep it moving. Reflexes keep parts of the gut informed about what's happening in other areas. When food enters the stomach, this triggers a reflex which speeds up the movement of the colon (the 'gastrocolic' reflex), emptying it ready for the next meal to come through. This is why small babies fill their

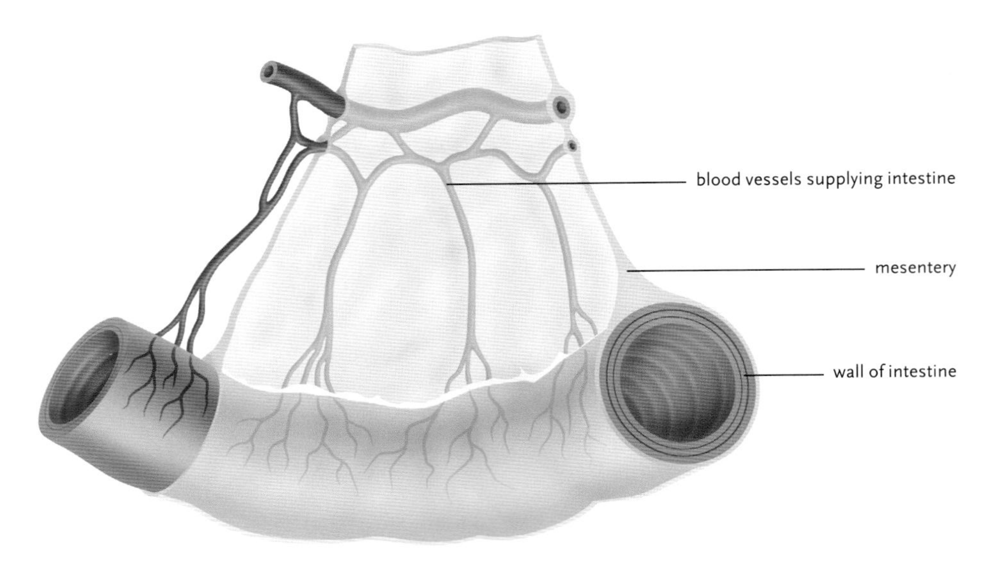

Rather than being held firmly in place, the small intestine is wrapped up and suspended by its mesentery; coils lying next to each other in the abdomen jostle for space as the food passes through.

nappies after feeding (luckily, we later learn to control this reflex). The gut, like the heart, also has autonomic nerves coming to it, and can be speeded up or slowed down by signals from these nerves.

If you were to open up the intestine, you would see clearly the circular ridges all along it, which help to increase the surface area available for absorption of nutrients into the blood. What you can't see with the naked eye are the smaller protrusions of the lining, which also increase the area for absorption. Looking through a microscope reveals that the tissue lining the intestine is like plush velvet; covered with minute projections called 'villi' (Latin for 'hairs'). If you could increase the magnification even more and look at a single cell, you would see even tinier hair-like structures: microvilli. The circular ridges, the villi and the microvilli are all designed to increase the surface area of the small intestine, maximizing the amount of useful nutrients you can absorb from the digested food in your gut.

After the trek through the metres and metres of small intestine, getting digested and releasing nutrients, what is left of your meal passes into the caecum – the beginning of the large intestine. In most animals, the caecum (Latin for 'blind') forms quite a pouch – a blind-ended, cul-de-sac of gut. Some animals use it as a place to ferment cellulose from plant cells: they nurture helpful bacteria there that can break down cellulose into sugars, which can then be absorbed. Horses do this – they have an enormous caecum, which stretches right up from their pelvis to their breastbone. We, on the other hand, don't do much in the way of cellulose digestion, and the human caecum isn't so impressive – it's just a slight pouch at the beginning of the colon. It sits low down in your belly on the right-hand side, tucked in near the bony knobble on the side of your pelvis (known to doctors by the grand title of the Anterior Superior Iliac Spine!)

Sticking out of the caecum is a bit of intestine that looks like a small earthworm has crawled in and nestled there. This is the vermiform (Latin for 'worm-like') appendix – known to its friends as, simply, the appendix. It is usually just a few centimetres long but its size can vary quite a bit. I helped at a particularly memorable operation to remove a troublesome appendix that I seem to remember was around 15 centimetres long. It trailed upwards

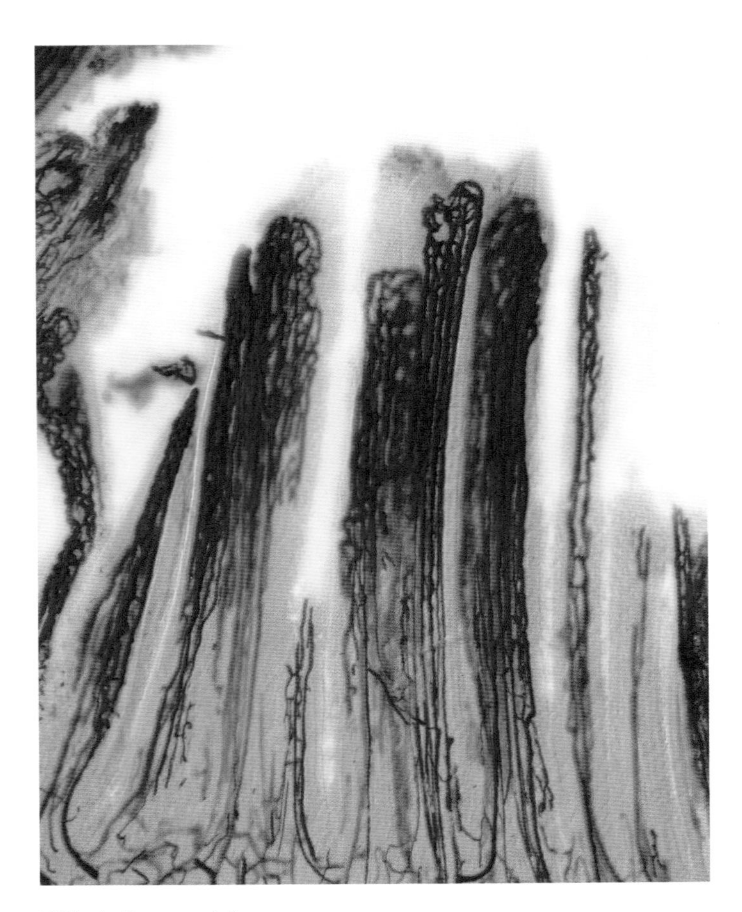

Villi in the ileum, specially stained to show up the blood vessels that absorb the products of digestion.

through the abdomen, finishing somewhere near the liver. The appendix does have its uses, despite what its detractors say: it's full of lymphoid tissue containing white blood cells, and is part of the immune system.

After the caecum comes the colon: the ascending colon runs up close to the liver, the transverse colon hangs down just under the stomach and finishes high up on the left, close to the spleen, and the descending colon runs down to the pelvis, where it empties into the rectum. Just as in the small intestine, muscle contractions in the colon push the digested material along it. Water is absorbed from its contents, which ends up as fairly solid faeces. This is very important – we would lose a lot of water if our colon didn't absorb it from the soon-to-be-faeces. When we have diarrhoea, we lose water and dehydrate very quickly and so have to drink a lot to make up for the loss of fluid from our bowels. The colon doesn't absorb many nutrients; the villi and microvilli that line the inside of the small intestine aren't found in the colon.

The rectum is about 12 centimetres long and quite stretchy. Contractions of the colon push faeces into the rectum, which fills up – and tells you it's filling up! But you can keep it all in until you're good and ready to let it out. At that point, the sphincters around the bottom of the rectum and the anal canal relax and open up, while the diaphragm and the muscles around the wall of your abdomen contract, squeezing the faeces along the four centimetres or so of the anal canal and out of the anus.

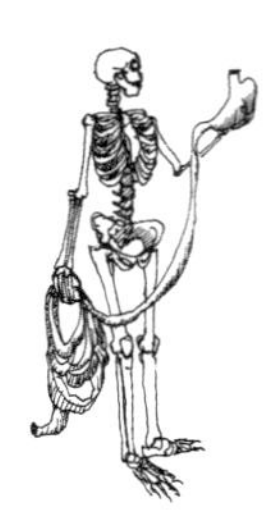

Common gut problems, and how to prevent them

We've seen how it all works when the stomach and intestines are healthy. I'll now turn to some common gastrointestinal ailments – once again, particularly those that are influenced by diet and lifestyle. First, I'll explore how different people might react to foods, food allergies (and the tests you can carry out for them) and food intolerance. Next, I'll look at peptic ulcers, stomach cancer, inflammatory bowel disease and colorectal carcinoma.

There's a lot of talk around at the moment about prebiotics and probiotics, so I'll have a go at untangling those. And finally, what about that trendiest of trendy therapies – colonic irrigation?

Food allergies

Food allergies involve an immune reaction to a particular type of food and are incredibly diverse in their manifestations. Someone who's allergic to wheat might get a nettle-rash type of reaction or a red rash from it, or they might start wheezing (sometimes called 'baker's asthma') or get a runny nose while cooking. Coeliac disease is a specific food allergy where someone is allergic to gluten, a protein in wheat.

Anaphylaxis: Anaphylaxis (Greek for 'guarding against') is a severe allergic reaction, involving the whole body, and can be extremely dangerous. For example, some people are so allergic to peanuts that the merest hint of peanut in a food can bring on anaphylaxis – an all-body allergic reaction. Some allergies cause local reactions, like the weal caused when you get stung by a nettle. Atopy (see pages 35–6) is a remote response to an allergen, causing a rash, runny nose or wheezing – responses that occur some distance away from the part of the body that originally came into contact with the allergen. Anaphylaxis takes it further: it's an extreme and total reaction where the entire body goes into a defensive downward spiral.

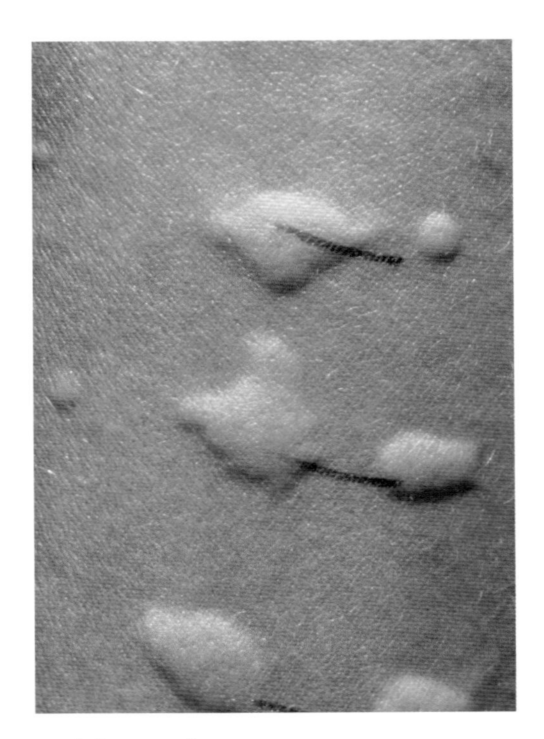

A real allergy test: allergens have been injected just under the skin, causing weals indicating an allergic response – but beware spurious tests like electrodermal skin testing.

The immune system, which usually works in a disciplined way to keep us protected from all manner of pathogens, is fooled by a small and often innocuous substance into thinking that the whole body is under deadly assault, and it brings out the big guns to fight it off. This is a design fault; life would be much simpler without anaphylaxis, but it seems to be the price we pay for an immune system that is constantly on stand-by and ready to mobilize. As someone goes into anaphylactic shock, their blood pressure drops, their heart rate rises, their face and throat swell up and they suffer nausea, vomiting and diarrhoea. It's life-threatening because of the swelling around the throat, which could choke the sufferer. Luckily, we have discovered ways of calming the immune system down: if someone goes into this potentially fatal state, they can be rescued with an injection of adrenaline, which damps down the immune reaction.

Having scared you all with the spectre of anaphylaxis, you can be reassured that the vast majority of us don't need to carry adrenaline around with us. Most allergic reactions are well short of the body-wide effects of anaphylaxis, appearing as rashes or wheezing. And although food allergies are on the increase (along with atopic diseases generally), they are still a lot less prevalent than most people think. About one in four people thinks they are allergic to some food or other. The real figures are much lower: one or two in every hundred adults and about six per cent of children. Be wary of spurious tests for food allergies, but if you are worried that you have an allergy – and this means experiencing symptoms more concrete than slight headaches or tiredness, such as swollen lips or eyes, rashes or wheezing after eating something – consult your doctor. A trained allergist will analyze your diet carefully, record the appearance of allergic symptoms when you have eaten potential allergens and carry out a skin-prick test.

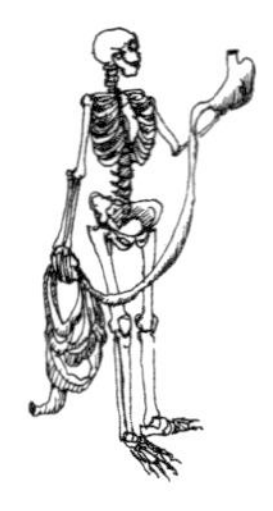

Just a few foods are responsible for nine-tenths of allergies; in adults, these include peanuts, hazelnuts, walnuts and seafood. In children, the most common food allergies are to egg, milk, peanuts, soy and wheat. Most children, happily, grow out of these allergies by the time they are three to five years old. People who develop allergies in adulthood are less lucky and can end up with a life-long allergy – but do remember that only one or two per cent of people have a real food allergy.

Cooking may reduce the potential of a food to cause allergies: fruit, vegetables, eggs and seafood are less allergenic when cooked than when raw. The type of cooking can also have an effect – frying or boiling peanuts may make them less liable to cause an immune reaction, while roasting them makes them worse. This may explain why peanut allergy is rare in China: the Chinese eat plenty of peanuts but not roasted ones.

Food allergies are due to a misdirection of the immune response – rather like the misplaced reaction to house dust mites and pollen that I discussed in The Lungs chapter. A healthy gut should be able to mount an immune response when it detects harmful bacteria or viruses but it should also be able to recognize non-harmful bacteria and leave them alone (essential for the flourishing of the gut's microflora or 'friendly bacteria'), and not get overheated by other innocuous substances in food.

The guts of infants have 'leakier' linings than adults', so more antigens (substances that may stimulate the immune response) can get through, and they have more food allergies. In adults, there are things which can temporarily increase the absorption of antigens from the gut, including alcohol, aspirin and exercise.

Breastfeeding and late weaning onto solid foods both seem to help prevent food allergies. It's recommended that babies at risk of developing atopic diseases (because of a family history) are breastfed for six to twelve months and not introduced to solid foods until they are at least five months old. If it's necessary to supplement or replace breast milk, hypoallergenic formula milk should be used, not soya milk – which can cause allergies. There's always a possibility that antigens from something the mother has eaten can end up in breast milk. However, there's not enough evidence to be sure that avoiding allergenic foods whilst pregnant or breastfeeding does anything to reduce the risk of a child developing atopic conditions.

If you have a food allergy, it can be quite hard to make sure you avoid your particular allergen. The complexities of food manufacturing and labelling make it difficult to be sure whether or not a particular manufactured food contains a particular antigen – another good reason to cook from scratch with ingredients you can identify. Managing children's diets can be a complete nightmare. What's in their school dinner? What if they share food with other children? This becomes a real issue if you have a child with a serious allergy but there are support groups and useful guidelines out there (see page 261).

Food intolerances

Food intolerances are quite different from food allergies – they don't involve an allergic reaction (although symptoms may be due to a general immune activation of blood cells) and they are much more common. You might experience windiness, bloating, stomach cramps and diarrhoea. This type of reaction is normally due to poor absorption of nutrients from the gut, which might happen if you are missing a particular enzyme (in the sense that you never had it in the first place), or it could be a reaction to chemicals – natural or artificial – in the food.

For example, lactose intolerance (the inability to digest milk) is caused by a genetic lack of the enzyme lactase, which breaks down lactose, the sugar in milk. The prevalence of this intolerance varies in different population groups. It affects almost everyone in Asian populations, 80 per cent of people of Afro-Caribbean descent and about 15 per cent of people of Northern European descent. Many more people think they have lactose intolerance than actually have it, so it's worth seeing your GP and getting properly tested if you suspect you're lacking in lactase. Being intolerant of lactose doesn't necessarily mean you have to eliminate dairy products from your diet; most people with this intolerance can still have up to 350 millilitres of milk per day without an unpleasant effect. You may find you can tolerate yoghurt more easily, as the milk is already fermented.

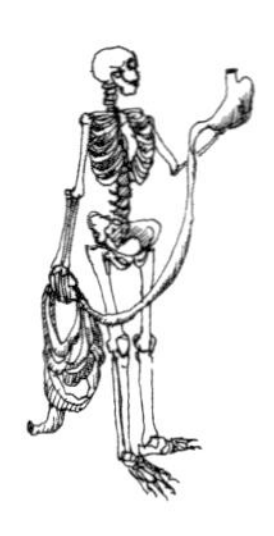

Peptic ulcers and stomach cancer

A peptic (Greek for 'digestive') ulcer is an open sore in the mucous membrane lining the digestive tract, usually in the stomach and duodenum. They can be very painful and may bleed.

For many years, the real cause of peptic ulcers was unknown. It was accepted that they could be brought on by stress and also, to some extent, from taking non-steroidal anti-inflammatory drugs (NSAIDs) like aspirin and ibuprofen. But it's now known that the main cause of peptic ulcers is a bacterial infection and it can be successfully treated with antibiotics. The bacterium in question is *Helicobacter pylori* – a spiral bacterium that likes to lurk around the pyloric sphincter.

But the diminutive *Helicobacter* doesn't just cause ulcers: in the 1990s, more and more evidence emerged to show that infection with this bug increased the risk of stomach cancer. However, rates of stomach cancer vary widely across the world. The rate in Yamagata in Japan is 13 times higher than it is in Los Angeles. Does this reflect different rates of *Helicobacter* infection in different countries or is there something else going on?

It seems that nitrates may be at fault. Nitrates get converted to nitrites in the stomach, especially if the stomach contents are less acid than they should be. These nitrites react with amino acids (the breakdown products of proteins) to make carcinogens. Vegetables are the main source of nitrates in our diet, so does this mean eating lots of vegetables might cause stomach cancer? No. Absolutely not. Eating lots of fresh fruit and vegetables *protects* you against stomach cancer – people who eat five to 20 servings of fruit and five to 20 servings of vegetables a week have half the risk of stomach cancer than those who eat hardly any. You have to look at the whole picture and study large populations to really see the effect of diet and lifestyle on health and disease. Nitrates are also used to cure meat and fish, and diets containing lots of dried or salted meat and fish (like the Japanese diet) seem to increase the risk of stomach cancer. Such diets have high levels of nitrates and also high levels of salt, which can directly damage the lining of the stomach.

It looks like diet and infection with *Helicobacter pylori* interact to give the widely varying rates of stomach cancer in different countries. Eating lots of

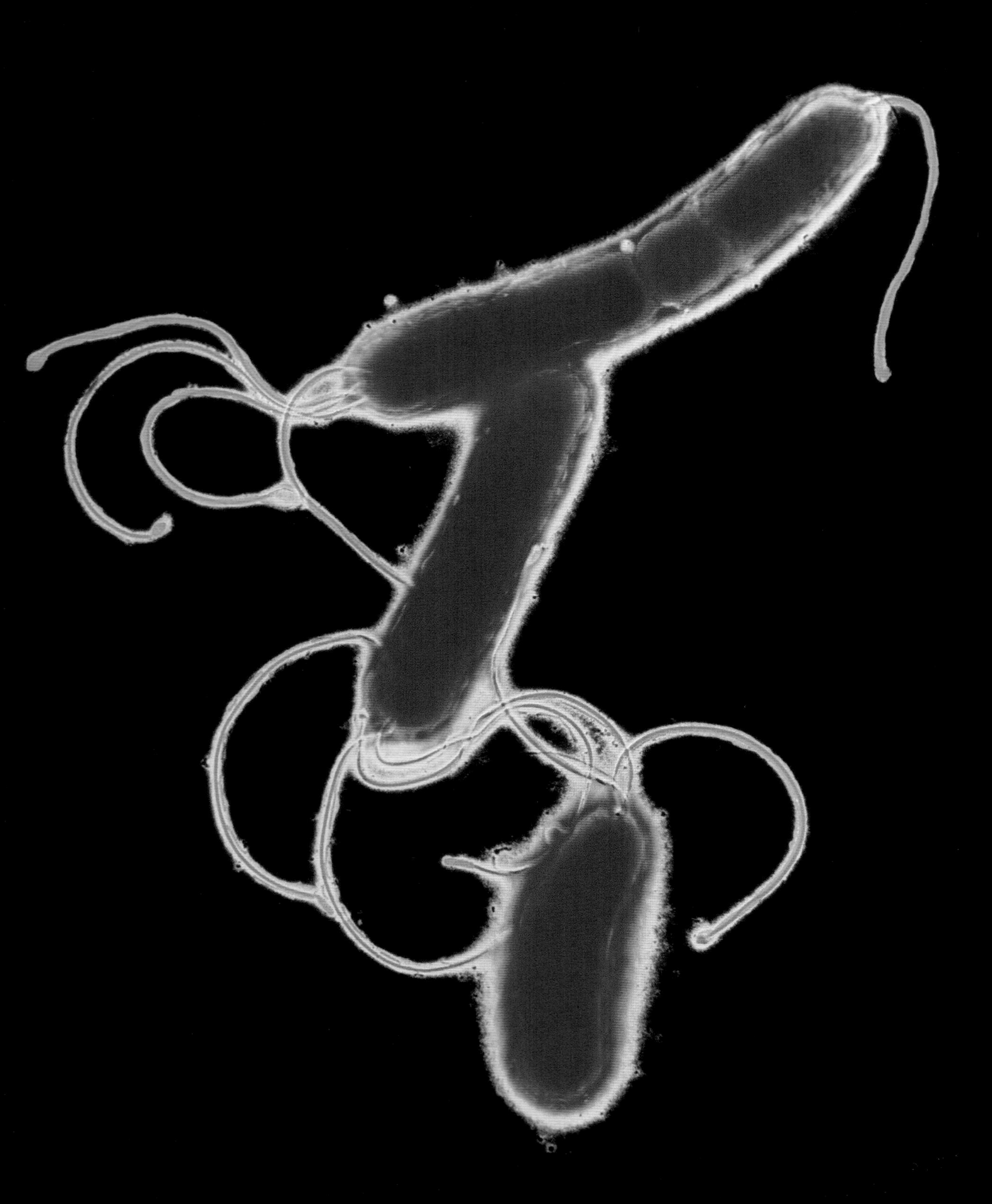

dried meat or fish leads to high levels of carcinogens in the stomach, a situation which can be made worse by inflammation due to salt and infection with *Helicobacter pylori*. Eating lots of fruit and vegetables, and reducing salt intake, combats the development of stomach cancer by reducing the inflammation and neutralizing the potential carcinogens.

NSAIDs: What about those NSAIDs? They do play a role in the development of ulcers. If you've had ulcers – and actually, even if you haven't – it makes sense to limit your use of NSAIDs. It's also extremely important to make sure you only take NSAID tablets on a full stomach. Too many people pop a couple of ibuprofen on an empty stomach, which is asking for trouble. NSAIDs are acidic and cause some direct damage to the lining of the stomach, but their main effect is indirect. They work to reduce inflammation by reducing prostaglandin production, but prostaglandins also 'turn on' mucus production in the stomach. NSAIDs lead to a reduction in the important protective layer of mucus, and leave the stomach lining exposed to its own acid. Aspirin has recently been found to be an important cause not only of ulcers but also of gastrointestinal perforation (holes that go right through the wall of the stomach and intestines).

Some foods, especially spicy foods, can be irritating to the stomach as well – if you have an ulcer, it's certainly a good idea to avoid foods like this that make your symptoms worse.

Inflammatory bowel disease

There are two main types of disease where the guts become inflamed: ulcerative colitis and Crohn's disease. The rates of both of these are high in developed countries, in particular in northern Europe and the US, and are rising in developing countries. Around 2.2 million people in Europe suffer from inflammatory bowel disease. The way that these diseases vary in different populations suggests that environmental factors play a large part. One important factor is smoking and, interestingly, another strong non-genetic factor is appendicectomy (removal of the

The stomach-ulcer-causing bacterium *Helicobacter pylori* has tentacle-like flagella (the green curls on this image) that it uses to propel itself along.

appendix). The relationship between inflammatory bowel disease and smoking is a bit odd: smoking seems to reduce the risk of ulcerative colitis but increase the risk of Crohn's disease. There's also a bit of a mystery with appendicectomy: it appears that people who've had their appendix surgically removed are at less risk of developing ulcerative colitis but at greater risk of getting Crohn's disease. Perhaps appendicitis itself (the inflammation of the appendix) protects against ulcerative colitis whilst predisposing someone to Crohn's. Perhaps the removal of the appendix has an effect, maybe altering the immunity of the gut in some way that makes ulcerative colitis more likely and Crohn's disease less likely.

As inflammatory bowel disease involves the immune system, it seems logical that antigens – from food or bacteria – might play some role. High sugar intake, chocolate and cola drinks have been identified as possible risk factors, while high levels of fruit, vegetables and fibre may be protective.

The picture becomes very muddied regarding childhood illnesses: some studies have suggested that children who have a lot of infections are more likely to get inflammatory bowel diseases but others seem to show a link with a low rate of childhood infection. Some people with Crohn's disease appear to have a *Mycobacterium* infection in their gut (a bacterium something like the one that causes tuberculosis) but tests of treatment with antibiotics have yielded equivocal results.

It seems likely that there are environmental factors contributing to the risk of these diseases – factors that we could modify, if only we could work out what they are. Plenty of people are beavering away, researching just that, so I suppose the advice has to be 'watch this space'. But do eat a healthy diet in the mean time.

Cancer of the colon and rectum

Colorectal cancer is the second most common cancer (after lung cancer) and is more common in developed than developing countries. Like most cancers, it's influenced by an interplay between genetic and environmental factors. Dietary factors include a diet rich in meat and fat, but lacking in fibre, calcium and folate (a B vitamin found particularly in green leafy

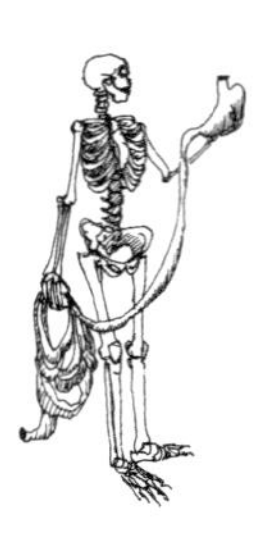

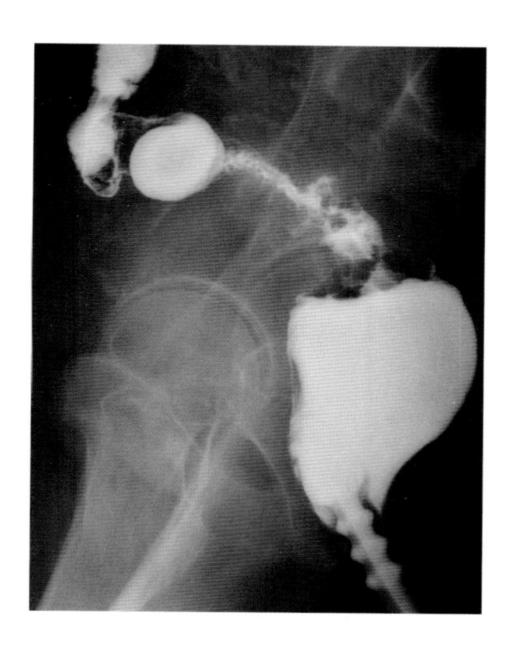

Colorectal cancer: the bowel has been filled with barium, showing up a part of it (the mottled orange area) narrowed by a cancerous growth.

vegetables). Other lifestyle factors include a sedentary lifestyle, obesity, insulin resistance and diabetes, smoking and high alcohol intake. People with ulcerative colitis and Crohn's disease also have an increased risk of colorectal cancer. Regular exercise lowers the risk of both developing and dying from colon cancer.

Cancers develop when mutations in genes cause cells to start acting in an uncontrolled way. It's as if a rule has been deleted and the cell forgets what it's meant to be doing and instead starts dividing to make a mass of cancer cells: a tumour or growth. In colorectal cancer, as in many others, the mutated genes are either ones that lead directly to runaway growth of cells, or put a spanner in the works of a cell's DNA repair mechanisms.

Diet plays a role as the source of potentially carcinogenic compounds that are either present in food or formed in the gut when food is digested. Carcinogenic compounds (called heterocyclic amines) are formed on the surface of meat when it's fried, grilled, roasted or barbecued. The gut flora make more carcinogenic compounds (N-nitroso compounds) when faced with a diet high in meat. Red meat (beef, pork and lamb) and processed meat (sausages, hamburgers, ham, bacon and canned meat) increase the risk of developing colon polyps (little bulges in the lining of the intestine) and colorectal cancer – and the risk increases the more meat you eat. Some bacteria, including probiotics (which we will explore further later), neutralize these carcinogens. Fermented fibre in the bowel also seems to stop cancer cells from thriving.

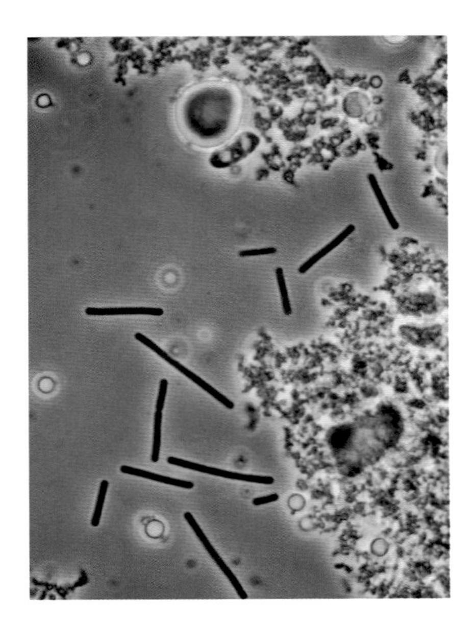

Living yoghurt: the thin black rods are *Lactobacillus* bacteria, magnified about 800 times. These tiny beasties are nothing to be afraid of: they are our allies in the fight against bacterial pathogens.

Gut microflora, diarrhoea and antibiotics

Three to five hundred different species of bacteria live in your guts. These little bugs are ten times more numerous than the cells of your body. It's quite a sobering thought: there are more of 'them' than there are of 'us'. We're a colony of bacteria in a human shell.

Just as you would speak about the native flora of a particular country, our bodies have their own native microflora – unique populations of bacteria living on us and in us. I love the word 'microflora' – it sounds so poetic and conjures up a wonderful mental image of a carpet of tiny flowers lining our intestines. The reality isn't quite so pretty, but it's still beautiful in a philosophical way: our guts play host to billions of bacteria, living with us in a symbiotic relationship. Symbiosis is much more than just putting up with each other: the two parties give each other mutual benefit. Your guts provide the bacteria with a nice home, plenty of food and not too many enemies to contend with. In return, you get some protection against nasty pathogenic bacteria, some nutrients from bacterial fermentation and vitamin K. The relationship is a little weighted in our favour – life just isn't fair, especially if you're a gut bacterium – however good a job they do in someone's gut, eventually their time is up and out they go. Up to 60 per cent of your faeces is made up of bacteria. The gut of a newborn baby is sterile, with no bacteria, but soon after birth, the lodgers move in and stay with us for life.

This relationship with your friendly gut bacteria isn't merely an added bonus to your health, it's absolutely fundamental to a healthy gut. If your gut microflora wilts, nasty bacteria – diarrhoeal pathogens like *Clostridium* and salmonella – can take a hold. (Continuing the floral theme, I suppose these malign bacteria would be evil weeds.) These bacteria cause diarrhoea, which leads to rapid dehydration and reduced absorption of nutrients. In developed countries, we take it for granted that a quick dose of antibiotics (in the floral analogy, weedkiller) will remove the pathogens and restore

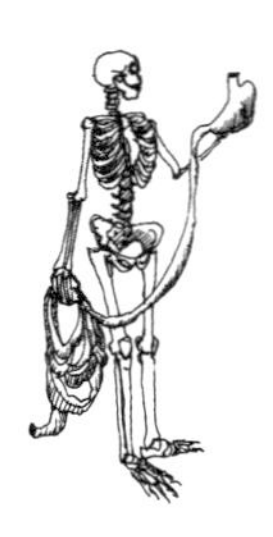

HEALTHY SCEPTICISM: COLONIC IRRIGATION

Cleaning the colon has been going on for at least 3,000 years – there are ancient Egyptian documents to prove it. Unfortunately, the evidence that it does anything for your health is, well, zilch. In an attempt to persuade me of its benefits, a well-meaning friend lent me a book. It was full of very technical and impressive-sounding pseudoscientific jargon but had no actual evidence to back up any of the claims it made about the many benefits of 'colon hydrotherapy'. Proponents speak of a fancy-sounding theory of 'autointoxication' – the body being poisoned by the build-up of toxins in a 'stagnant' bowel. This theory was popular in the nineteenth century but discredited early in the twentieth, so as an argument, it's slightly out of date! Not only is colonic irrigation of no proven benefit but the hazards are serious, ranging from physiological disturbances to heart failure and perforation of the gut. Practitioners using poorly sterilized equipment have spread serious infections. And even if you were to escape these more serious effects, you'd presumably be left with a bowel cleaned of its useful bacteria and more vulnerable to infection.

us to health. We tend to forget that diarrhoea is a deadly serious condition that kills thousands of people every year in developing countries. The solution can also be part of the problem: broad-spectrum antibiotics can wipe out our gut microflora as well as the pathogens, leaving it defenceless, which is why people often get diarrhoea after taking antibiotics for other infections.

As prevention is always better than cure, it seems sensible to try to reduce the risk of diarrhoea in the first place. In developed countries, we are lucky to have clean water and sanitation. General hygiene, including simple things like hand-washing plays a very important role.

Probiotics and prebiotics

Probiotics: A review in *The Lancet* described probiotics as 'live organisms that . . . beneficially affect the host by improving the properties of indigenous microflora, hampering the growth of diarrhoeal pathogens and boosting . . . immunity'. More succinctly, probiotics are 'friendly bacteria'. There's real interest in using probiotics to fight diarrhoea in developing countries, where sanitation is poor and a vaccine is as yet unavailable. Evidence shows that friendly bacteria seem to boost gut microflora and reduce the risk of diarrhoea.

Diarrhoea is still a problem in developed countries, especially antibiotic-associated diarrhoea. Probiotics may help to restore the gut microflora after it has been indiscriminately wiped out by antibiotics. It's not clear if all probiotic preparations manage to deliver their friendly bacteria into the gut (remember the vat of stomach acid along the way), but nevertheless, probiotics seem to be generally good for the health of your guts, with the potential to compete with pathogenic bacteria and produce lactic acid (which the pathogens hate), promote the growth of your own symbiotic gut microflora, stimulate the immune system and neutralize dietary carcinogens. You can buy probiotics in capsules but there are similar bacteria in live yoghurts and yoghurt drinks (but some of these are very sugary) – check they contain strains like *Lactobacillus acidophilus* or species of *Bifidobacterium*. An added advantage of live yoghurt is that its proteins and calcium also stimulate the immune system and are good for gut health. Live yoghurt may be a way for people with lactose intolerance to get calcium into their diet: the live bacteria digest lactose and can eliminate symptoms of intolerance.

Prebiotics: Prebiotics are another group of supplements that, rather than containing bacteria themselves, are food molecules that friendly bacteria will enjoy: pet food, for your internal pets. They are snacks that keep lactobacilli and bifidobacteria happy, including sugars and other organic compounds found naturally in foods such as wheat, onions, garlic, leeks, artichokes and bananas. The evidence for their efficacy is, as yet, scanty, but that doesn't stop the pill-pushers trying to sell you trendy prebiotic supplements – remember that the same compounds are available in fruit and vegetables.

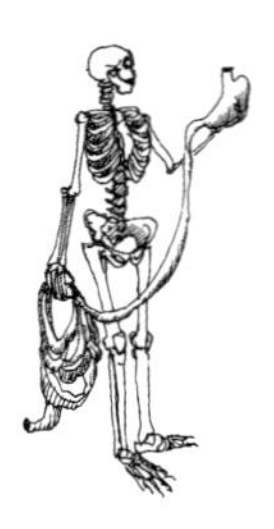

Five ways to keep your stomach and intestines healthy:

- Eat a diet rich in fruit, vegetables and fibre and cut down on salt, dried meat and fish. This will keep your guts healthy and reduce your risk of cancer.

- Try to keep your weight down – obesity is linked to cancer and can also cause acid reflux (when the acidic contents of the stomach gets pushed back up into the gullet).

- Avoid foods that give you heartburn.

- Reduce your risk of stomach and gut infections by thoroughly washing fruit and vegetables. If you're somewhere with a dodgy water supply, drink bottled water and be careful about eating uncooked foods like salads.

- Never take aspirin or ibuprofen (NSAIDs) on an empty stomach.

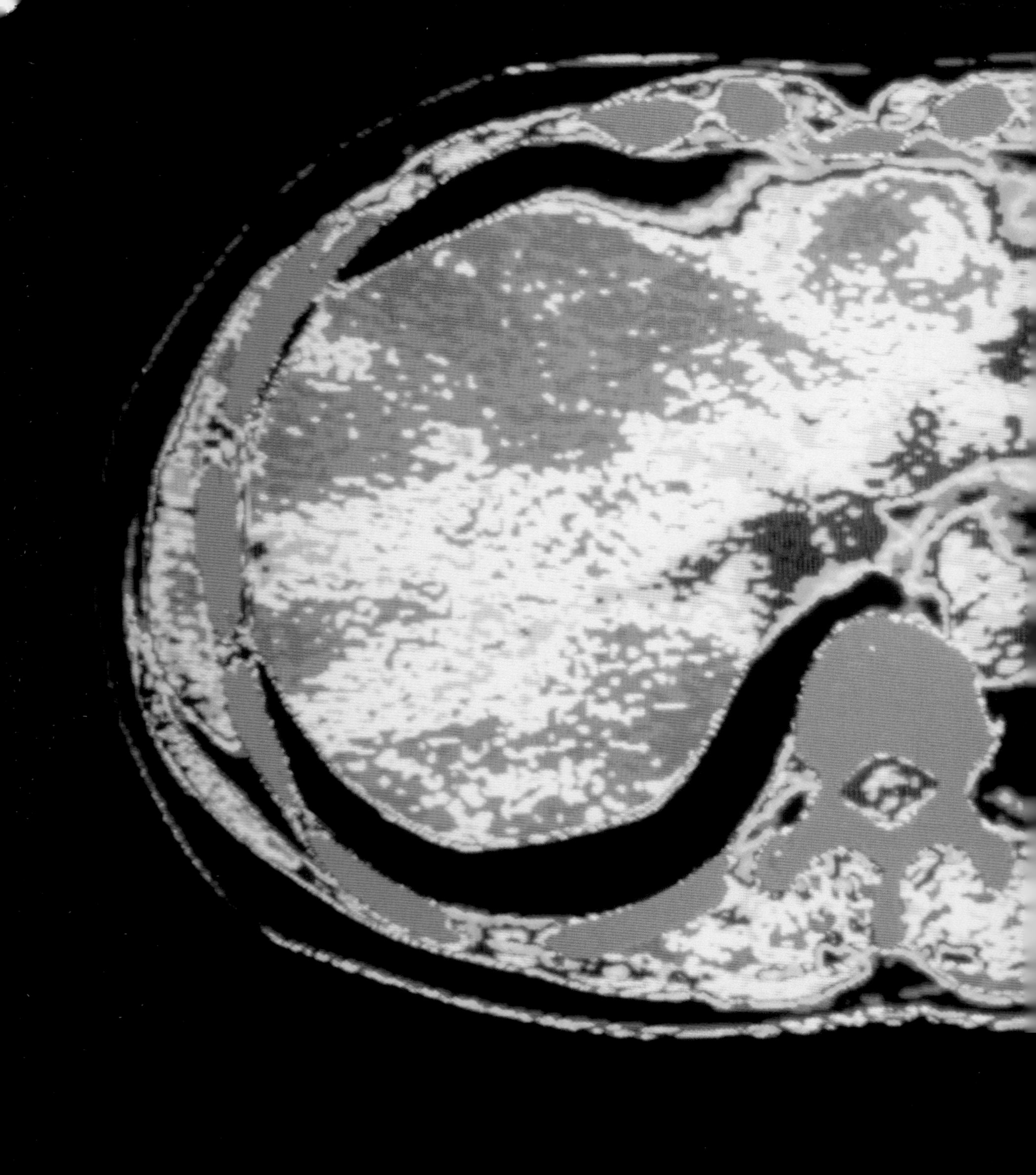

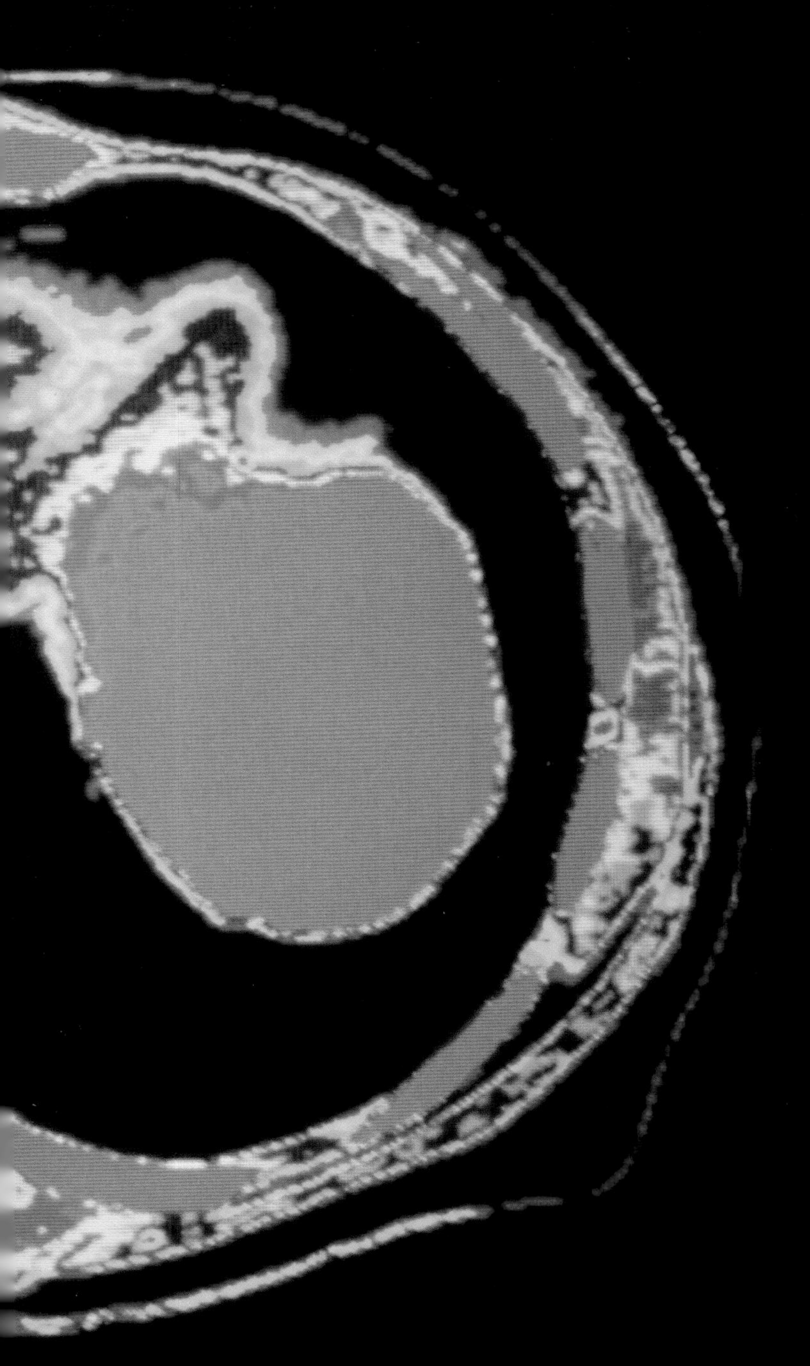

FOUR

THE LIVER

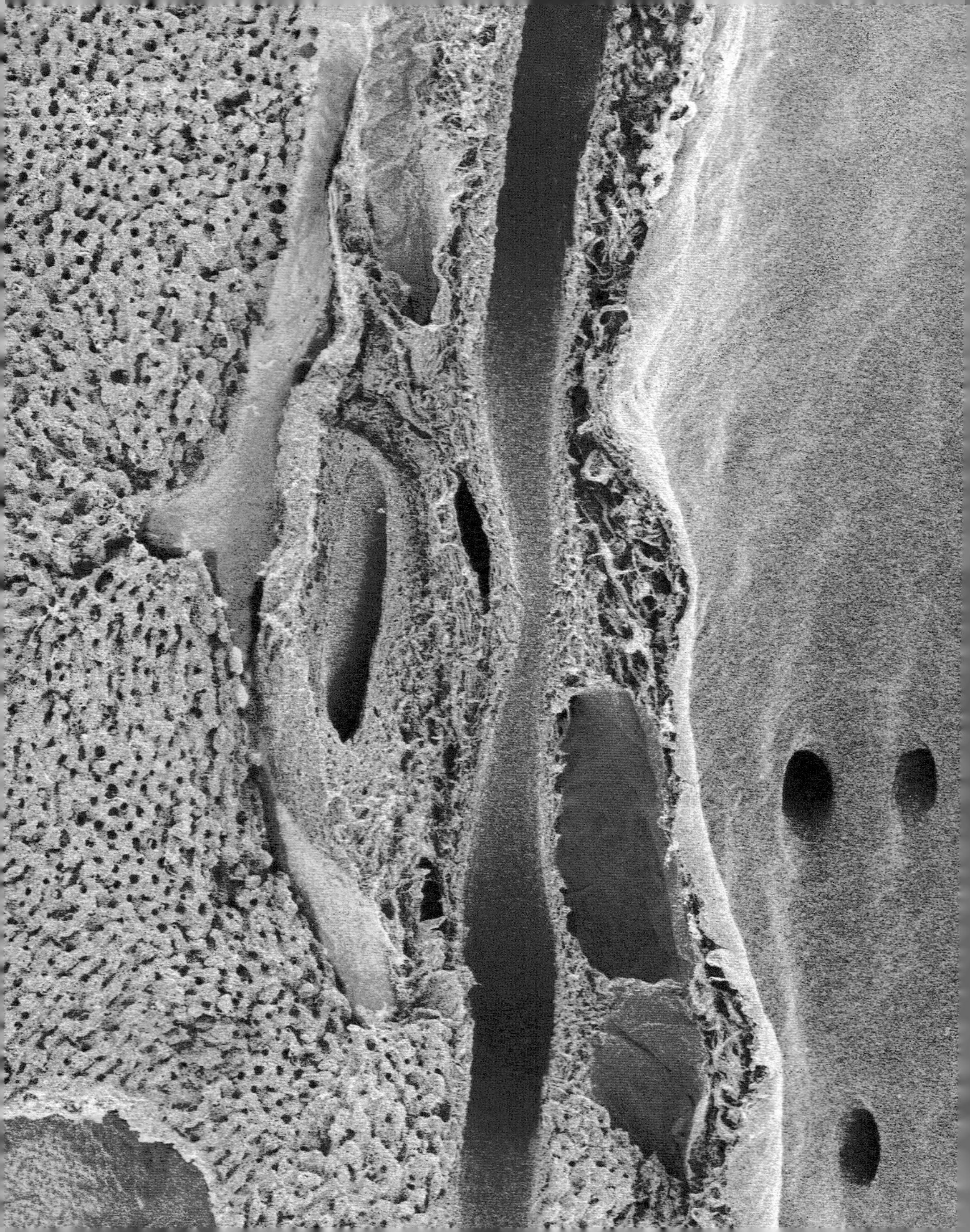

The liver is the biggest organ in the body (not counting the skin) weighing between 1.5 and 3 kilograms. It's a fantastic multi-tasker, juggling hundreds of different jobs. One of its main jobs is making and breaking proteins, carbohydrates and lipids (fats). It gets its raw materials from the gut, taking in all the nutrients and deciding what to do with them according to the body's needs: it acts like a factory, recycling plant, storage depot and waste disposal facility all rolled into one. It can make brand-new glucose and lipids, cholesterol and important proteins used in blood clotting.

The liver also plays an important part in digestion, producing bile, which is stored in the gall-bladder and used in the intestine to help the gut absorb fats from food. It has an important role in detoxification and deactivation, breaking down drugs and alcohol, and it also gets rid of tired, old red blood cells. It plays a role in the immune system, protecting the body from infection. The liver does so many things that it's not surprising that the consequences of liver disease are serious; the body cannot survive without a functioning liver.

Previous pages **A slice of liver: this coloured CT scan shows a horizontal section through the lower ribcage, showing the liver (yellow and orange) and the stomach (round and pink).**

Left **Service ducts: this bit of liver magnified under an electron microscope shows the arteries (red), bile ducts (purple) and portal vein (orange, on the right).**

Where is the liver?

The liver is a very large organ, lying tucked up underneath the ribs on the right-hand side. The diaphragm domes up into the chest, and the liver sits up against the diaphragm on the right, while the stomach lies just below it on the left. Like a lot of other organs, the liver is held in shape by the structures around it – take it out of the abdominal cavity and it flops and spreads out on the table. If you've ever prepared liver to eat, you'll know what it feels like: human liver is very similar to any other animal liver in texture and it's a similar reddish-brown. The liver has two main lobes, right and left, which are separated in the front by a mesentery-like membrane, the falciform ligament, attaching the liver to the inside of the wall of the abdomen.

What does your liver look like?

The liver is one of those organs that doesn't reveal its secrets when you cut into it. You don't see much detail with the naked eye: it just looks like a dense brownish-red tissue. But pick it up and turn it over and then you see some clues.

A pear-shaped bag, the gall-bladder, hangs underneath the liver. There's a large artery, the hepatic artery, which brings oxygenated blood to the liver and, close to it, a large vein (the portal vein) disappears into the organ. A greenish tube leaves the liver at about the same point: this carries bile out of the liver. This area on the under-surface of the liver, with tubes and vessels going in and out, is called the 'porta hepatis' (Latin for 'gateway of the liver').

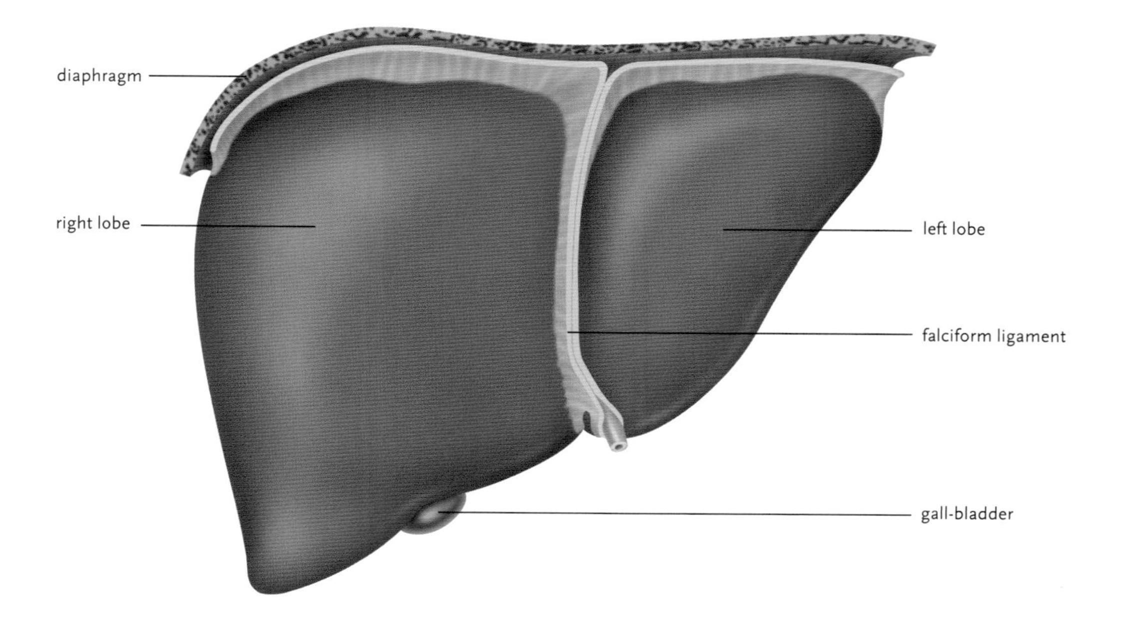

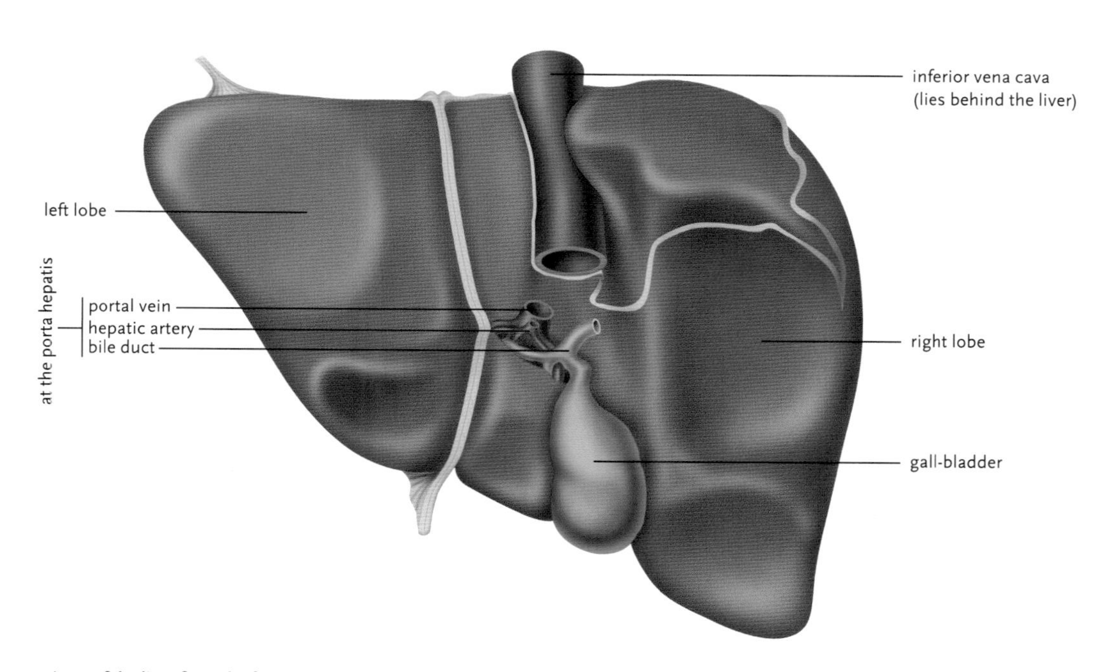

Views of the liver from the front (above) and from the back (below).

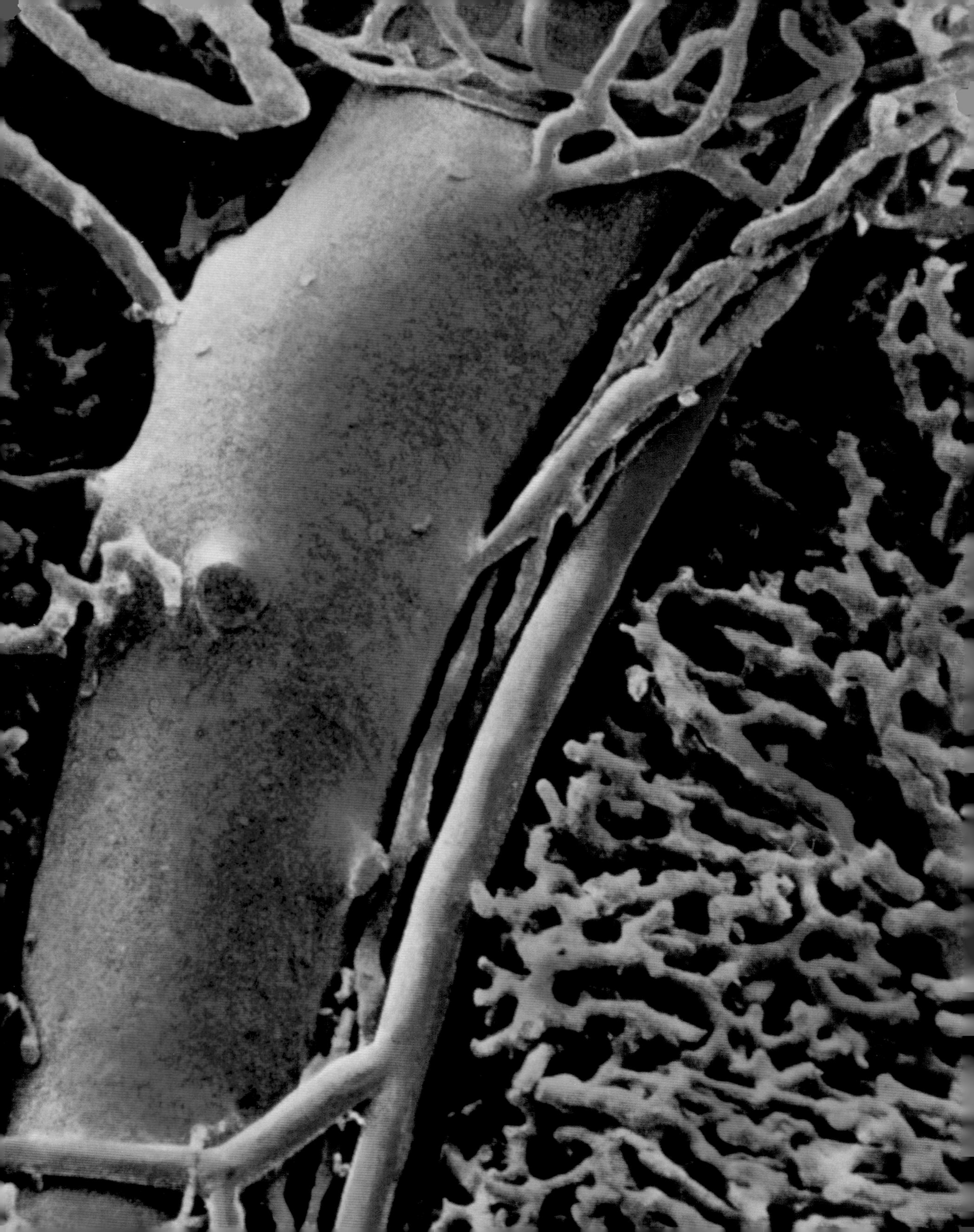

How the liver works

The portal vein brings the products of digestion (the nutrients absorbed from the gut into the bloodstream) to be sorted out by the liver. The flow of blood in the portal vein increases after you've eaten a meal. Contained in this nutrient-rich blood are the building blocks the body needs to make vital proteins, carbohydrates and lipids (fats and oils).

In the liver, amino acids, the building blocks of protein, are either joined up into new proteins or, if they're not needed, broken down to make urea, which the kidneys get rid of in urine. The liver takes molecules of glucose and joins them together into a large molecule, glycogen (the animal equivalent to the starch made by plants), for storage. When the body needs the energy in this sugar store, the liver breaks down glycogen back into glucose. It's like having a sugar loaf that you can chip bits off when you need to. (As well as doling out this stored glucose, the liver has a very neat trick of being able to make brand new glucose when it's needed.)

The liver also builds up and breaks down lipids. Oil and water don't mix, and the liver does the very important job of packaging oil into globules so that it can be transported in blood. Because it gets the blood from the guts, the liver is also well-placed to act as a warden; checking to make sure that no bacteria have made it into the body. The liver contains cells of the immune system, which recognize foreign bodies and gobble them up.

The portal vein also carries toxins to the liver. It's the liver's job to remove them before they can harm any other part of the body, either by breaking them down into safer chemicals or packing them up into non-toxic substances that can safely re-enter the blood and get excreted in the urine by the kidneys. Alcohol and other drugs are inactivated in a similar way. And the liver also breaks down and deactivates hormones. These are the chemical messengers of the body, such as adrenaline, which helps get the heart beating more quickly when you're in a stressful situation. It wouldn't do you any good if hormones circulated in the bloodstream indefinitely – or else their effect would just go on and on, so the liver takes on the job of breaking them down.

Loaded with nutrients: the great trunk of the portal vein (blue) sends branches into the liver tissue. The hepatic artery (red) lies at its side.

Another of the liver's jobs is to break down old red blood cells, in particular breaking down haemoglobin into bilirubin. The liver gets rid of bilirubin by putting it into bile, which passes into the gut and leaves the body in the faeces. But bile is more than just waste; it contains substances that are useful to the intestines: bile acids (which the liver makes from cholesterol). These are detergents and act like washing-up liquid; they break large globs of fat down into smaller globs (just as washing-up liquid breaks fat into smaller globs that you can wash off the plate). The small globs can then be broken down into their constituents and absorbed.

The liver makes bile constantly – but humans don't tend to eat continuously – so the liver has a bag to store bile in until it's needed: the gall-bladder. (Some animals, like rats, have no gall-bladder as they eat almost continuously so don't need to store bile between meals.) The bile made in the liver is collected by tiny vessels called bile canaliculi (Latin for 'tiny

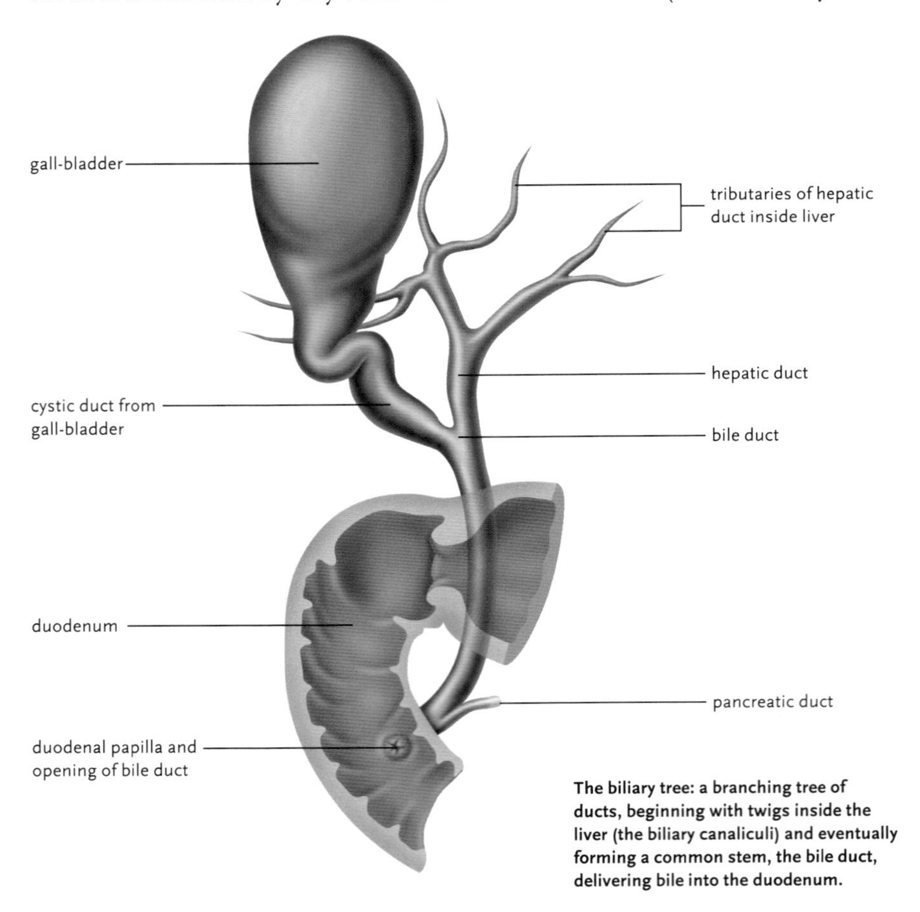

The biliary tree: a branching tree of ducts, beginning with twigs inside the liver (the biliary canaliculi) and eventually forming a common stem, the bile duct, delivering bile into the duodenum.

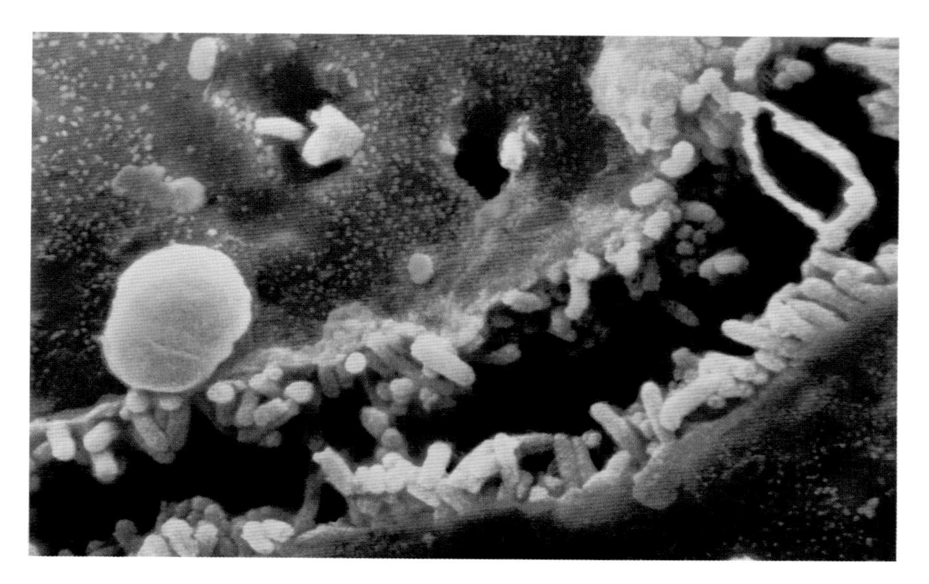

A tiny duct or canaliculus (green) inside the liver collects the bile manufactured by the hepatocytes; magnified around 30,000 times.

canals'), which eventually join together into a trunk. (This structure looks rather like a tree and, indeed, is often referred to as the biliary tree.) The duct that forms the trunk of this tree leaves the liver and has a side branch going off to the gall-bladder, so that bile can be diverted into this holding bay when it's not needed. Then, when you eat a meal, chemical signals from the stomach tell the gall-bladder to contract and the luminous green bile is pushed down the bile duct and squirts into the duodenum, together with enzyme-rich secretions from the pancreas, which opens into the duodenum at the same point.

Incidentally, the end of the bile duct and the circle of muscle around it have great names: the 'ampulla of Vater' and the 'sphincter of Oddi'. A bit like astronomers naming stars after themselves, many anatomical bits and pieces bear the names of the people who first recorded them. There's a move away from these eponymous names today: the ampulla of Vater is officially called the hepatopancreatic ampulla. I think this is a shame; the names are wonderful, making the body sound like a fantastic, Lord of the Rings landscape and it's sad that the people who found these things won't be commemorated any more.

Common liver problems, and how to prevent them

The most prevalent disease of the liver is utterly lifestyle related: alcoholic liver disease. Other common problems, like viral hepatitis and gallstones, are also affected by dietary and lifestyle factors, so we'll take a brief look at those as well.

Alcoholic liver disease

The most common cause of an unhealthy liver is excessive alcohol consumption. When you drink alcohol, it is absorbed from the gut into your blood. Alcohol, like many other substances that we humans positively enjoy and seek out, is a toxin: being drunk is being poisoned. It's quite upsetting to think that a delicious glass of chilled Sancerre is a poisoned chalice. Happily, the liver is there to protect us, and a healthy liver can comfortably eliminate about one unit of alcoholic poison an hour. If you're drinking more quickly than this, it struggles and you start to feel drunk.

Alcohol has a range of damaging effects on the liver, from fatty changes to hepatitis and cirrhosis. The liver makes fat as it detoxifies alcohol, and if you drink too much alcohol, the cells of your liver become swollen with the fat the liver makes from it, producing a rather interesting 'Swiss cheese' effect when you look at the liver through the microscope. This isn't damage as such and disappears if you cut down on your drinking.

Alcoholic hepatitis is inflammation of the liver. White blood cells move into the tissue of the liver and liver cells die. If you don't stop drinking at this stage, cirrhosis is pretty much inevitable. Cirrhosis is a strange disease: parts of the liver are scarred and useless whilst little nodules of liver tissue are trying to regenerate. At first, the liver may grow larger, but as the scarring progresses, the liver shrinks.

Long-standing liver disease has widespread effects. Perhaps the most well-known and obvious effect is jaundice: yellowing of the whites of the eyes and of the skin. It's caused by a backlog of the breakdown product of red blood cells – bilirubin – which is normally excreted by the liver into the

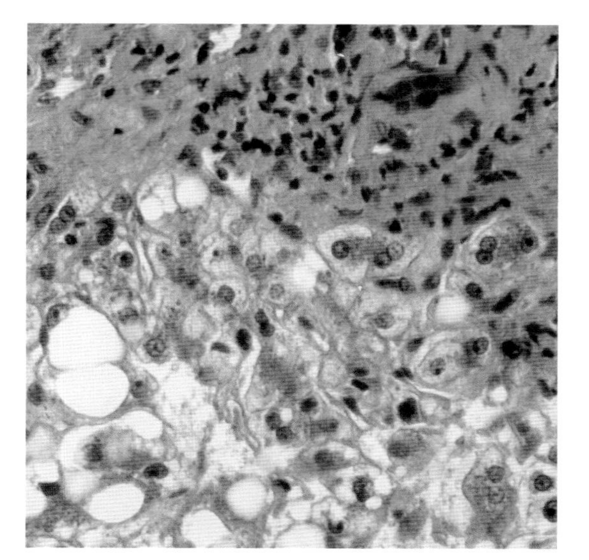

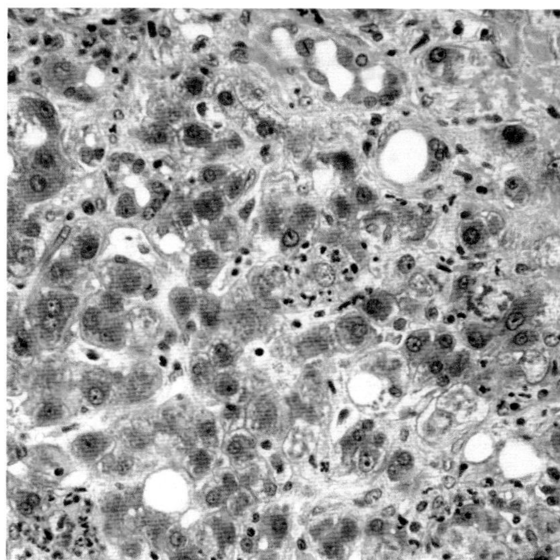

Left **The 'Swiss cheese' appearance of a fatty liver (the 'holes' are fat droplets inside the cells).**
Right **The little black granules inside the liver cells are a sign of alcoholic hepatitis.**

bile. This build-up can cause itching as well as the yellowing colour. Because the liver breaks down hormones, an unhealthy liver can lead to unusual levels of hormones: men may develop breasts, while their testicles shrink. Little spider-like blood vessels appear on the upper chest, and the palms of the hands become flushed. The legs swell with tissue fluid and the belly becomes swollen and full of fluid. If the liver isn't doing its job of detoxifying, poisonous substances start to build up in the blood. These toxins affect the brain, leading to drowsiness, confusion and disorientation. In severe and life-threatening liver disease, these toxins can lead to coma and death.

Should I stop drinking? Does this mean we shouldn't drink any alcohol at all? No, not really. The liver is quite up to the job of detoxifying a couple of glasses of wine – and we know that a moderate amount of alcohol can have positive effects on the cardiovascular system. The key to a healthy liver is moderation. Look after your liver and it will rise to your aid when you fancy a pint of beer or a glass of wine. But don't take it for granted and don't overload it. You need your liver for all sorts of essential jobs and replacing it

GIN LANE.

Gin cursed Fiend, with Fury fraught,
Makes human Race a Prey.
It enters by a deadly Draught,
And steals our Life away.

Virtue and Truth, driv'n to Despair,
Its Rage compells to fly,
But cherishes with hellish Care,
Theft, Murder, Perjury.

Damn'd Cup! that on the Vitals preys,
That liquid Fire contains
Which Madness to the Heart conveys,
And rolls it thro' the Veins.

isn't easy. Drinking a couple of units of alcohol per day seems to have positive effects on the health of your heart, but more than this and you're increasing your risk of cancer and cirrhosis. There's a strong association between increasing consumption and increasing rates of cirrhosis.

Researchers into alcoholic liver disease used to remark on the low rates in the UK – but no longer. The number of people dying from alcohol-related disease has risen 18 per cent since 2000. The amount we drank in Britain doubled between 1960 and 2002, as alcohol got cheaper, was more readily available round the clock and was more heavily advertised. The rate of cirrhosis rose as well: part of a wider trend across Northern Europe – we share a beer-drinking, spirit-drinking culture.

Some 40 years ago, the UK government identified two interventions that had been successful in reducing the national burden of alcoholic liver disease: keeping the price of alcohol high and restricting its availability, by keeping a tight rein on the opening hours of places selling alcohol. Today's government, though often castigated for being a nanny state, seems to shy away from measures that might have a real impact on disease prevention and public health when it comes to alcohol (and smoking). We have seen deregulation of the controls that used to be in place and the new licensing law for England and Wales actually states that conditions related 'solely to the health of the customers' are firmly outside its domain. And although the Advertising Standards Agency has banned adverts for alcopops, it doesn't stop them being readily available.

So, it really is up to us as individuals. The Department of Health's guidelines advise men not to drink more than three to four units (up to two pints of ordinary strength lager or four glasses of wine) per day and women not to drink more than two to three units (1 to 1.5 pints of lager or three glasses of wine) per day. If you drink more than this, your risk of liver disease really starts to mount up – and the risk of liver disease increases with each extra drink.

Most of the time, the amount of alcohol you drink only has an effect on your own health. Apart from blokes (and increasingly women as well, it seems) getting horrendously inebriated and starting fights, the other time

In Hogarth's *Gin Lane*, spirits are seen as one of the social evils of the eighteenth century – and they are once again threatening the nation's health.

when alcohol can affect the health of someone other than the person drinking it is during pregnancy. Drinking whilst pregnant can affect the physical and mental development of the fetus – and binge-drinking is particularly damaging. The most fragile time of embryonic development is during the first one to two months, so there's a possibility of harm befalling the embryo before a woman actually knows that she is pregnant. To cut out this risk completely, the best thing to do is to stop drinking alcohol altogether if you are trying to conceive, rather than making that change after finding out about being pregnant.

HEALTHY SCEPTICISM: HANGOVER CURES

Bad news, I'm afraid. A comprehensive review in the BMJ in 2005 found there was no real evidence that any hangover cure, conventional or alternative, complementary or over-the-counter, has any effect at all on preventing or treating a hangover! Apparently, two billion pounds are lost in Britain each year because of hangovers – mostly through sickness absence. Shame on us!

Viral hepatitis

A few viruses specifically infect the liver and some can cause serious damage. There are no specific treatments – apart from measures which can ease the symptoms – but these are preventable infections.

Hepatitis A virus infects the liver, where it multiplies, gets into the bile, into the gut and is then excreted in faeces. This means it spreads rapidly in places where sanitation is poor and drinking water is contaminated with sewage. Shellfish that have lived in contaminated water can carry the virus. So, you should try to avoid potentially contaminated water or shellfish. If you have to drink water you're unsure about, boiling it for ten minutes

kills off hepatitis A. Infection with the hepatitis A virus may cause a general illness, and some people recover after this stage – but the infection can persist and go on to cause hepatitis, with jaundice and an enlarged liver. Although most people recover from hepatitis within six weeks, it can occasionally turn into a severe and fatal disease.

Hepatitis B virus is either spread via blood, through blood transfusions or sharing needles, or through unprotected sex, because the virus also gets into semen. The first stage of the infection sometimes goes unnoticed, or appears as a general illness similar to hepatitis A. Although most people recover completely, some go on to have long-term liver disease. In the UK, people such as health workers, who are at risk of hepatitis B infection because of their job, are vaccinated against the virus.

As well as other varieties of hepatitis, there are other viruses that occasionally cause liver disease. These include the Epstein-Barr virus (which causes glandular fever) and cytomegalovirus.

Gallstones

It's unusual to have gallstones – literally, stones in the gall-bladder – before you are 30 years old. They're more common in women than in men. Overall, about one in ten adults in the UK has them, rising to about one in four people over the age of 60.

Gallstones vary considerably in size and shape: some are round and smooth, others are knobbly, and some are faceted. They are mostly made of cholesterol, although they can also contain bilirubin (from the breakdown of haemoglobin) and calcium. It's perfectly normal for cholesterol to be in bile but it will only crystallize into stones if there is so much dissolved that the bile can't hold any more – and is 'supersaturated'. A 'lazy' gall-bladder, which doesn't contract much can also lead to gallstone formation. Cholesterol gallstones are pearly white. Gallstones containing bilirubin are brown or black and tend to form in the gall-bladders of people with blood disorders, because large numbers of red blood cells are being broken down. This type of stone may also be caused by a bacterial infection of the gall-bladder, which is likely to happen when something blocks the normal flow of bile.

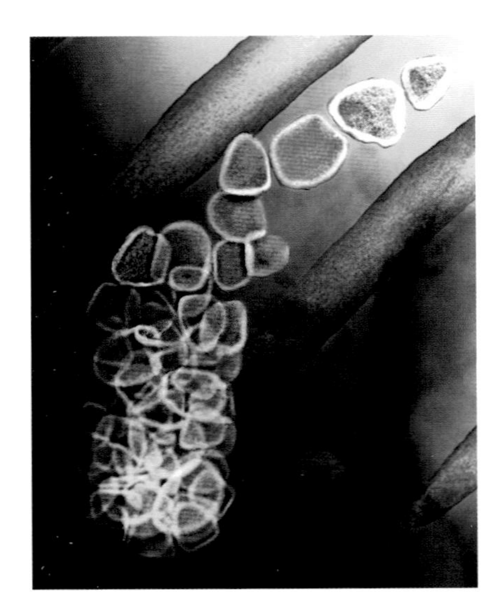

An X-ray shows up calcium-containing gallstones (coloured orange here) crammed inside the gall-bladder.

If they contain calcium, gallstones may be seen on X-rays of the abdomen – but an ultrasound scan is the best way of seeing most gallstones. The vast majority of gallstones don't cause any problems or symptoms at all but some can be very troublesome, occasionally getting stuck in the bile duct and causing severe pain just under the ribs on the right side. This 'biliary colic' may come on after a particularly over-indulgent or fatty meal. If a gallstone gets wedged in the bile duct, completely blocking it, the gall-bladder swells up with bile and becomes inflamed. This makes the lining of the gall-bladder more susceptible to infection (a condition called cholecystitis). This infection can get very nasty and the gall-bladder can rupture, pouring pus and bile into the abdominal cavity – an emergency which requires swift surgery. Gallstones that manage to pass down the bile duct can also block the pancreatic duct, causing pancreatitis. The general rule is that, if gallstones are causing symptoms, then they should be removed, and nowadays, this can usually be done by keyhole (laparoscopic) surgery.

When I was at medical school, it was a well-known axiom that patients likely to have gallstones were 'fair, fat, female and over forty'. As well as sounding a bit derogatory, this turns out to be wrong. Women of a certain age, weight and complexion are not especially singled out. Having said that, women are affected more than men, and women who have had children or take the contraceptive pill seem to be more at risk. If people in your family have had gallstones, you will also have a higher risk.

The risk factors relevant to gallstone development are very similar to those in atheroma: an unhealthy diet (high in calories, low in fibre and high in refined carbohydrates) and a sedentary lifestyle both increase your risk. Obesity is a risk factor in itself – but so is crash-dieting and rapid weight loss. Having diabetes increases the risk but, rather strangely, drinking coffee seems to lower your risk of having problematic gallstones.

Five ways to keep your liver and gall-bladder healthy:

- Moderate how much alcohol you drink: you are at risk of alcoholic liver disease if you are a woman drinking more than two units per day or a man drinking more than three.
- Practise safe sex: hepatitis can be spread through sexual intercourse.
- Eat a healthy diet: eating a diet high in fibre and low in refined sugar will lower your risk of gallstones.
- Keep your weight down – obesity increases the risk of gallstones – but don't crash diet; losing weight too quickly is also a risk factor.
- Stay active: people with a sedentary lifestyle are more likely to develop gallstones, whereas people who play sport regularly are less likely to get them.

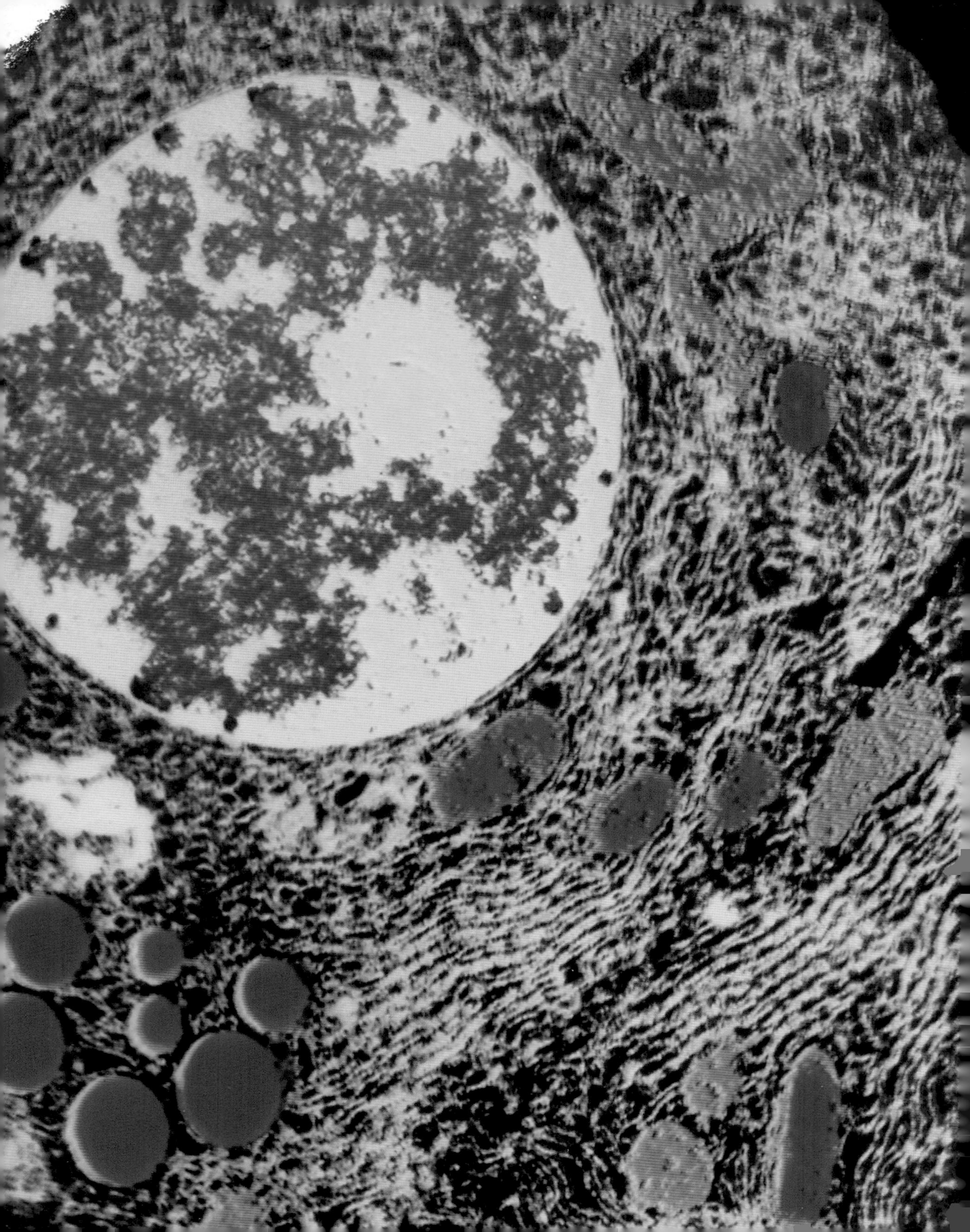

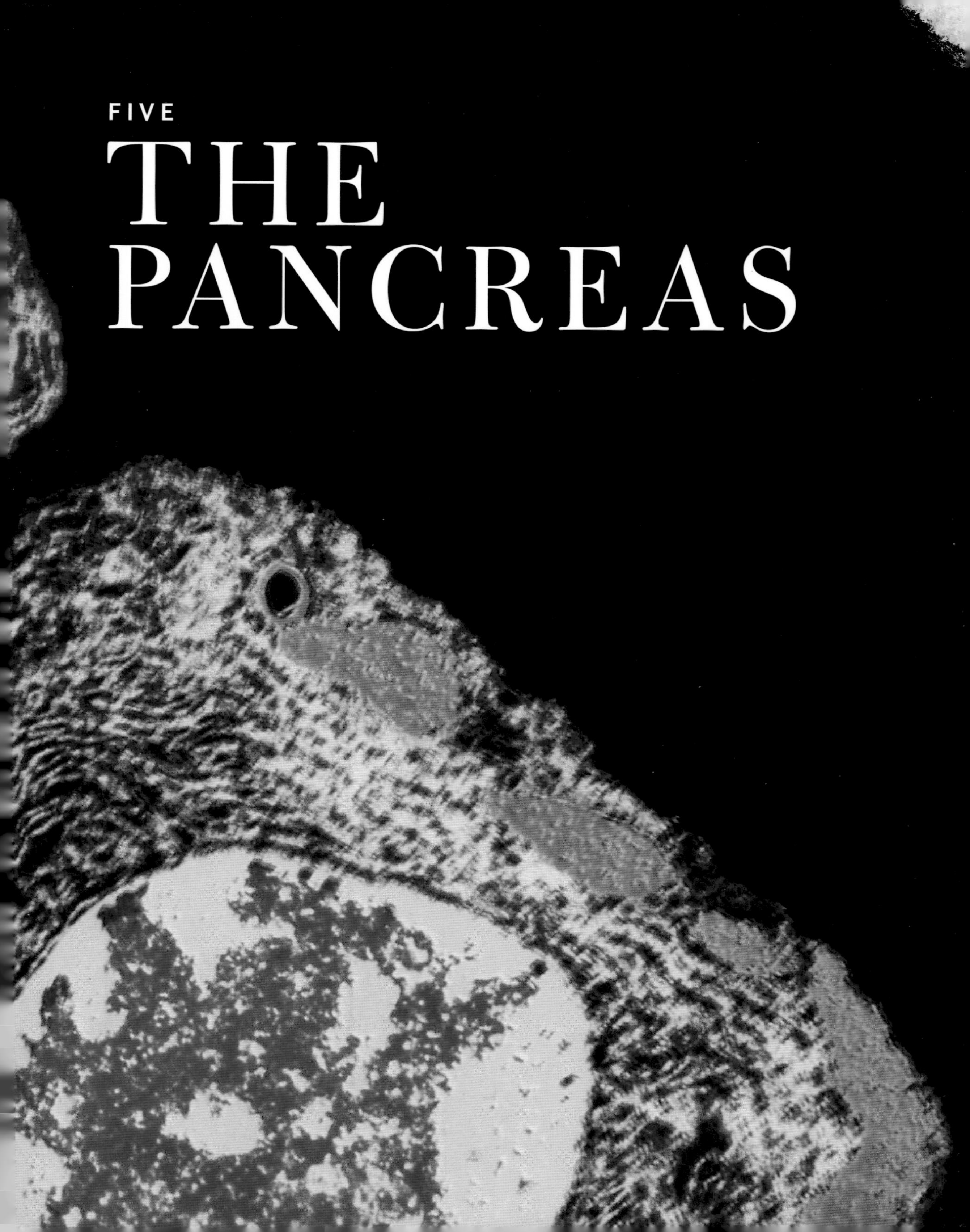

FIVE

THE PANCREAS

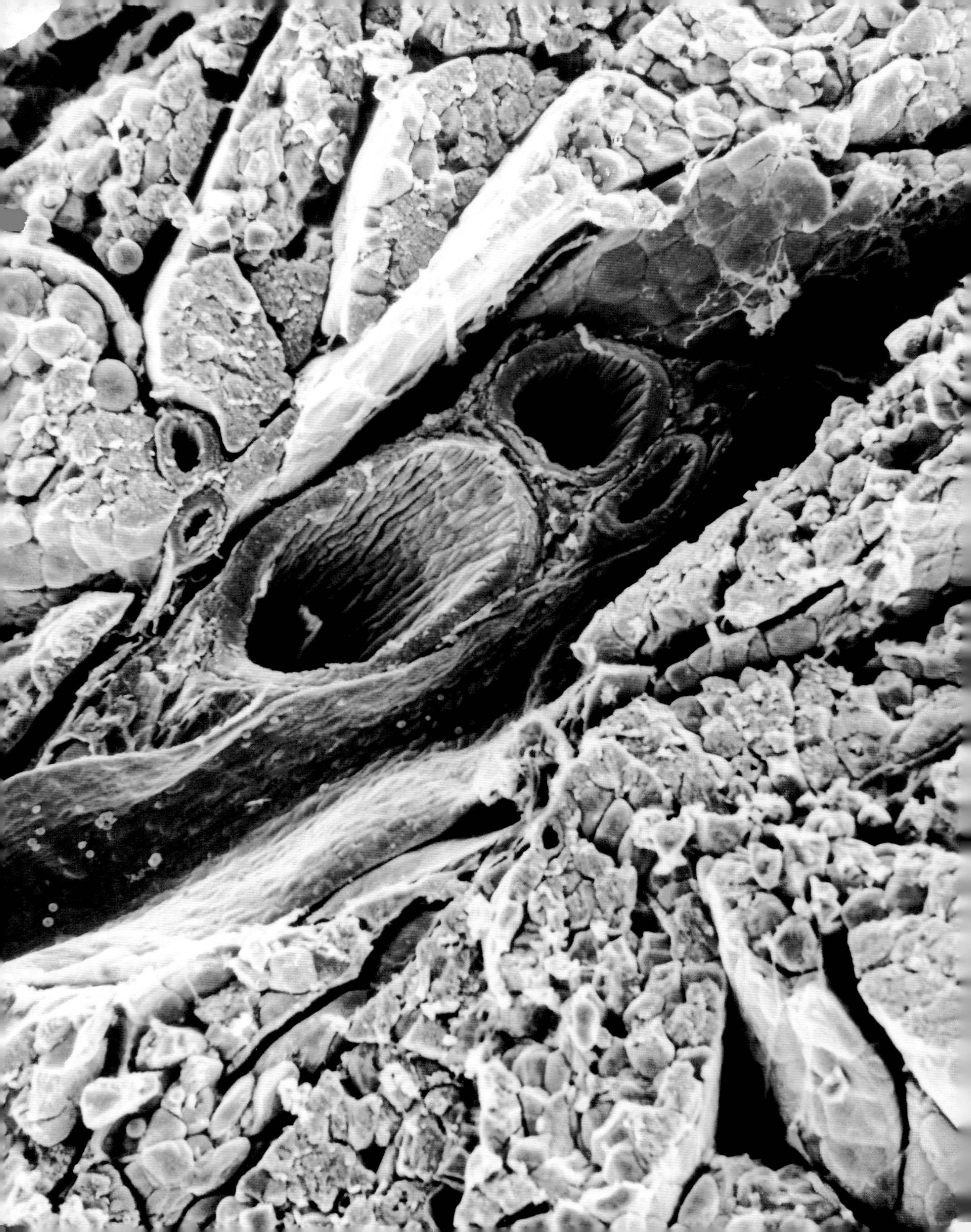

The pancreas is a long, thin, leaf-shaped gland that sits under the liver and behind the stomach. It's a particularly special type of gland because it works in two ways: it makes hormones, which go into the bloodstream and it makes pancreatic juice, which goes into the intestines.

The main hormones the pancreas produces are insulin and glucagon, which play a vital role in keeping the amount of the sugar in the blood within a narrow range. Too much blood sugar damages the lining of your blood vessels; too little starves the brain of energy. Pancreatic juice contains alkali to neutralize stomach acid and digestive enzymes to break down proteins, carbohydrates and fats. This is secreted into the pancreatic duct, which joins up with the bile duct just before it opens into the duodenum. If the pancreatic duct becomes blocked, the enzyme-rich digestive juice can't flow out into the gut, and builds up inside the pancreas.

Previous pages **Seen up close under an electron microscope, these pancreatic cells are full of packets of digestive enzymes (orange), ready for delivery to the intestines.**

Left **Aiding digestion: cells arranged in clusters around the ducts of the pancreas that carry the digestive enzymes away to the duodenum.**

Where is the pancreas?

Anatomists describe the pancreas rather as though it were a small animal, with a head, neck, body and tail. Its head is tucked into the C-shaped curve of the duodenum and its body goes off to the left, behind the stomach. The tip of its tail touches the spleen, up under the ribs on the left-hand side.

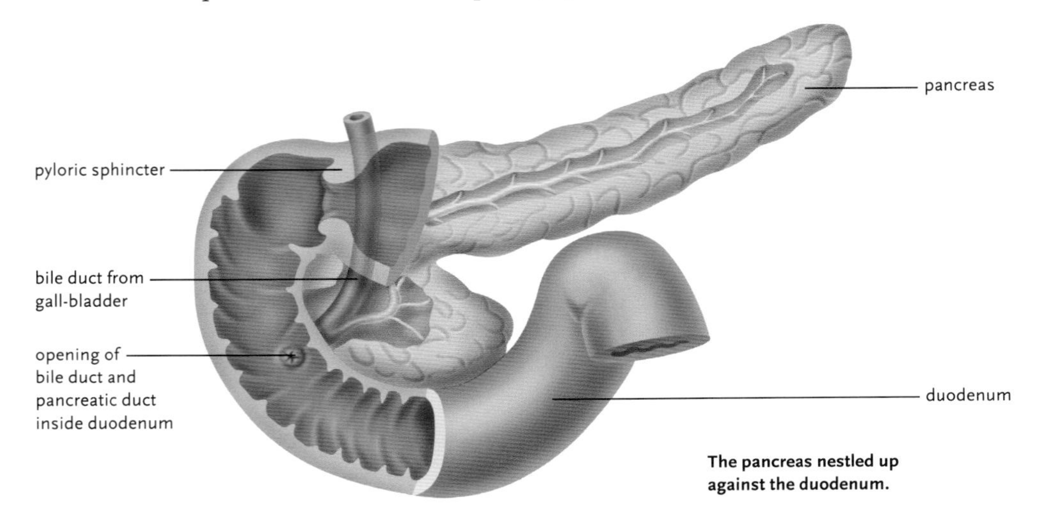

The pancreas nestled up against the duodenum.

What does your pancreas look like?

The surface of the pancreas has the typical, bobbly appearance of glandular tissue. Through a microscope, you can see the cells making the hormones, arranged in little nests, called pancreatic islets or the islets of Langerhans. Beta cells in the islets make insulin, whilst alpha cells make glucagon. These are the two main hormones made by the pancreas but there are many others. In between the hormone-secreting islets are the cells that secrete the digestive juices. These are also arranged in little clusters, opening into small ducts which eventually join together to form the main pancreatic duct. The clusters of cells are called 'acini'. The main pancreatic duct is like the stem of a vine, with branches and twigs dripping with bunches of grapes: the acinar cells.

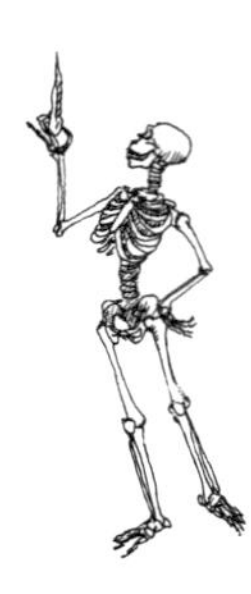

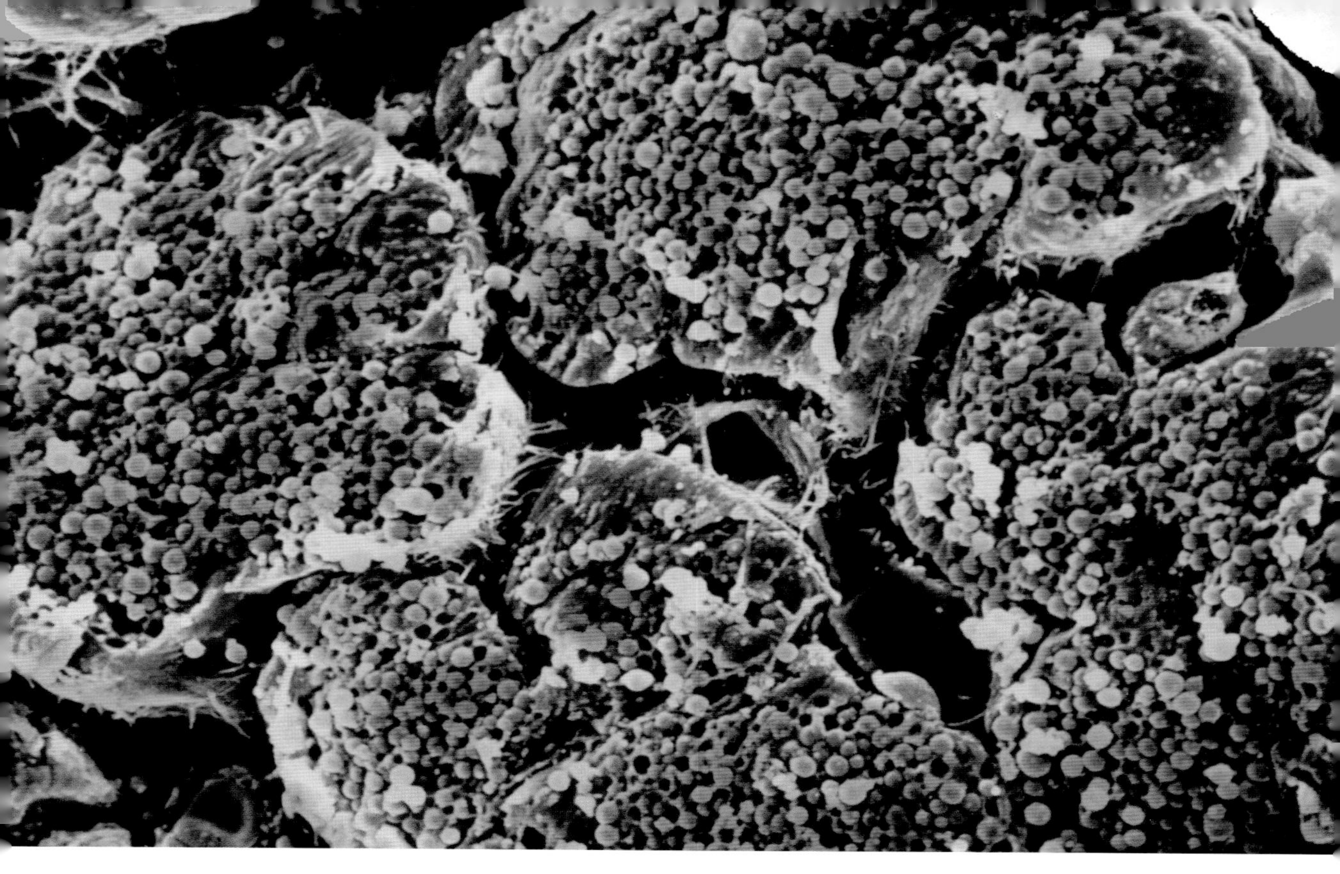
Grapes on a vine: clusters of acinar cells (purple).

How the pancreas works

It's extremely important that the levels of glucose in your blood stay within a narrow range. The liver empties glucose into your blood, mostly from breaking down its glycogen stores, although it can manufacture brand-new glucose if it's needed. Glucose is taken out of the blood and used as fuel by all the cells in the body. The greediest cells are those in your brain – out of the 200 grams of glucose used by the body each day, 100 grams is fuel that is used for the brain.

Between meals, your insulin levels are low, but after a meal, the pancreatic islets pump out insulin to bring down the level of glucose in the blood. Insulin encourages tissues like fat and muscle to suck up more glucose and, at the same time, tells the liver to stop making new glucose. Glucagon, also produced in the pancreas, has the opposite effect: raising blood sugar levels. Another hormone that raises blood sugar is adrenaline, which comes from the adrenal glands. Adrenaline is produced in response to stress, as part of the 'flight or fight' response; it makes sure that the brain is alert and well-supplied with glucose.

The other function of the pancreas is all about the digestion of food. The pancreatic duct collects the pancreatic juice made by the acini, and carries it to the duodenum. Bile and pancreatic juices are released together into the gut when a meal is on its way. How do they know when to start? Together, a clever series of nerve reflexes and hormones cause the pancreas and gall-bladder to squeeze their respective contents into the duodenum. The mere sight and smell of food gets the brain excited, causing it to send messages to the pancreas and gall-bladder via the vagus (Latin for 'vagrant' or 'wanderer') nerves. This pair of nerves begins in the brainstem, then they wander all the way down the neck into the thorax and end up in the abdomen.

The vagus nerves carry autonomic nerve fibres – the nerves that look after the things that go on in the body automatically. You don't have to think 'Ah – food. Pancreas and gall-bladder – are you ready?' The vagus nerves look after that, so you can get on with enjoying your gustatory experience without having to orchestrate your guts consciously to digest the imminent meal. The vagus nerves also respond to stretching of the stomach as it fills up with food and relay the message on to the pancreas and gall-bladder.

Hormones are also involved. As food enters the small intestine, two hormones are released from the cells lining the gut: cholecystokinin (Greek for 'gall-bladder-mover') stimulates the gall-bladder to release bile, and secretin causes the pancreatic acinar cells to give up their enzyme-rich juices. The enzymes in the juice include amylase, which breaks starch down into sugars; lipase, which digests lipids (fats), and proteases, which break down proteins into amino acids. The pancreas also secretes bicarbonate, which neutralizes the stomach acid that has been mixed with the food. As the digested food passes down the duodenum, through the jejunum and into the ileum, other hormones tell the pancreas to stop secreting its juices.

Common pancreas problems, and how to prevent them

The major lifestyle and diet-related disease of the pancreas is type 2 diabetes mellitus. Other pancreatic diseases are also influenced by lifestyle and diet, including pancreatitis.

Diabetes

If your pancreas doesn't produce enough insulin or if the tissues of your body are resisting its message, then levels of glucose in the blood may get too high. This condition, called hyperglycaemia, makes you feel tired and thirsty, and makes you need to urinate often. The kidneys don't normally get rid of glucose but if blood levels are too high, it will be excreted in your urine. This gives us the name of the disease: diabetes is Greek for 'passing through' and mellitus is Latin for 'honey'. The urine becomes sweet. In order to get rid of this glucose, the kidneys produce a larger volume of urine than normal, which can lead to dangerous dehydration (because the water has to come from somewhere) and blood clots. In the long term, high levels of glucose damage blood vessels.

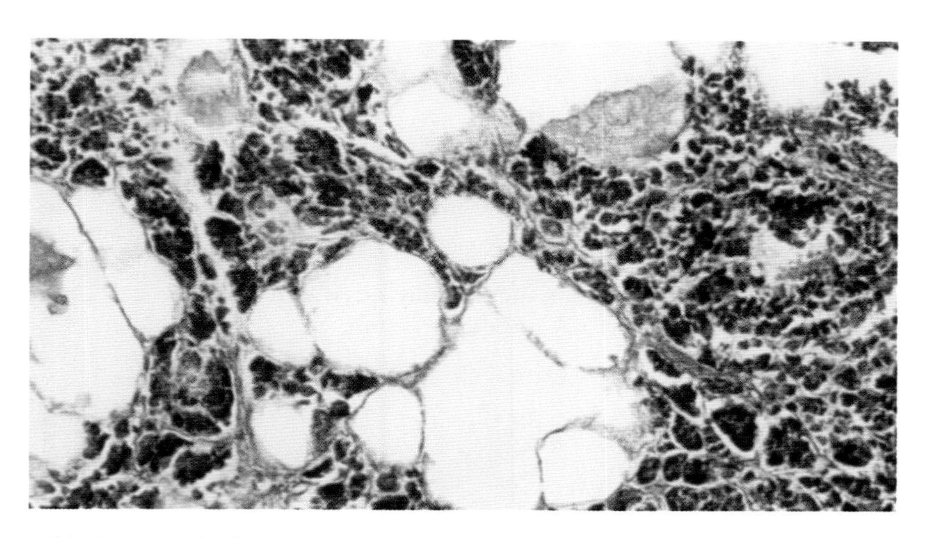

A diabetic pancreas: the islets of Langerhans have disappeared, leaving white spaces, while the acinar cells (purple) are unaffected.

The hormones glucagon and adrenaline raise blood sugar levels when the body is burning glucose quickly, but if the level of glucose in the blood falls too low (hypoglycaemia), glucagon and adrenaline can struggle. To bring glucose levels up, your adrenal glands pump out lots of adrenaline and these high levels cause paleness, sweating, a pounding, racing heart and trembling; symptoms that insulin-dependent diabetics and their families quickly learn to recognize. If the glucose levels don't rise and the glucose-hungry brain cells are deprived, the person starts to slip from consciousness and may then fall into a coma.

Type 1 diabetes: There are two types of diabetes, cunningly named diabetes type 1 and diabetes type 2. Type 1, or insulin-dependent diabetes, is an auto-immune disease, in which the immune system attacks the pancreas and destroys the cells that produce insulin. Sufferers from this disease need to inject insulin. There are both genetic and environmental factors in the development of type 1 diabetes. While we know that some people have a genetic susceptibility to the disease, we don't really understand the environmental factors very well. It's possible that viral infections play a part and a clean environment in childhood may also have an effect (the same 'hygiene hypothesis' that is put forward to explain the increase in atopic and allergic conditions that we looked at in The Lungs chapter).

Type 2 diabetes: The other, more common type of diabetes (type 2 – which used to be known as non-insulin-dependent diabetes) also develops in a mixture of genetic and environmental factors. This type of diabetes is on the increase; it's linked to poor diet, sedentary lifestyle and obesity. Type 2 diabetes used to be known as 'adult-onset diabetes' but not any more: type 2 diabetes is on the rise amongst children in the UK and in other developed countries. So is childhood obesity. This has increased substantially in developed countries in the last 20 years: about eight per cent of adolescents in the UK are clinically obese and another 15 per cent are seriously overweight. Obesity in adolescence causes immediate health problems and sets the stage for ill health later in life, in particular, type 2 diabetes.

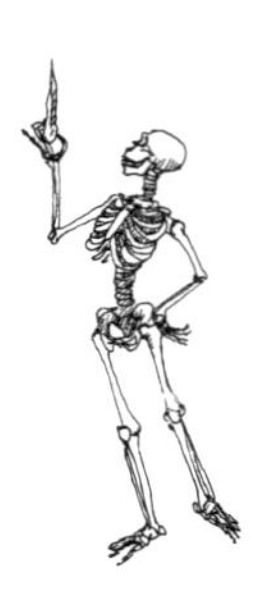

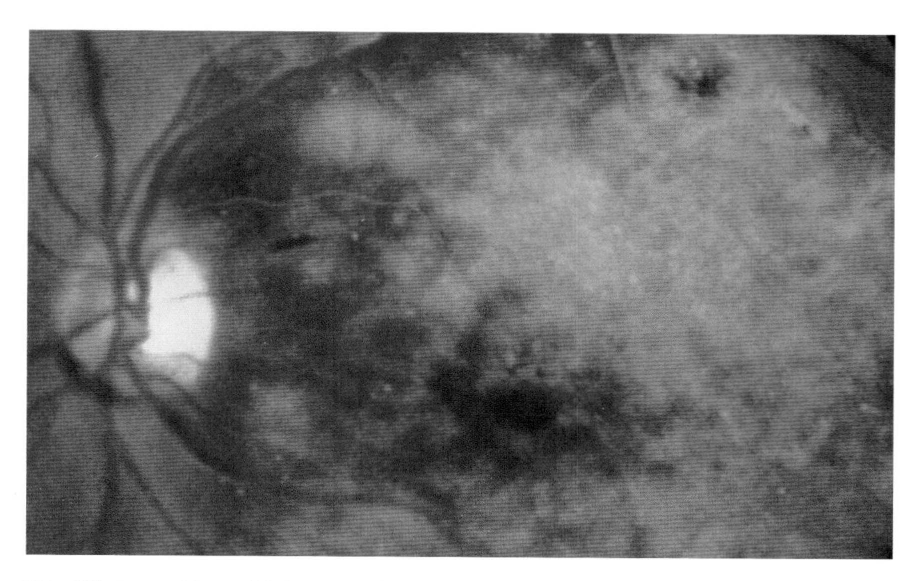

Risk of blindness: with an ophthalmoscope, damage caused to the retina by diabetes can be clearly seen in the patchy pink area on the right (the yellow area on the left is the optic disc).

In type 2 diabetes, the main problem is neither the pancreas nor the amount of insulin it makes, but that the tissues of your body become resistant to its message. In contrast with type 1, insulin levels may be high in type 2 diabetes but the effect of even those increased levels of insulin is reduced. Insulin resistance seems to develop due to a sugar-rich diet and sedentary lifestyle: your tissues become less sensitive to insulin, putting the pancreas under pressure to produce more insulin than normal to bring blood sugar levels down. People who have a sedentary lifestyle are more resistant (or less sensitive) to insulin than those who are physically active.

Type 2 diabetes seems to be one of the health problems that can arise when the body experiences a diet and lifestyle far removed from those that kept prehistoric hunter-gatherers healthy. The development of insulin resistance might seem like a design fault but it makes sense from an evolutionary point of view: it could be something which evolved to help our ancestors survive famine and may still be advantageous in countries where famine remains a part of life. However, in the affluent west, it is an adaptation that leaves us vulnerable to the high fat and high energy diet that we can easily hunt and gather.

More than sugar levels: Diabetes has a big knock-on effect on other organs. It wreaks havoc with nerves and small blood vessels, causing damage to the retina of the eye (which can lead to blindness), and to the kidneys (which can lead to kidney failure). It can also cause ulcers of the feet (which can lead to gangrene and amputation), erectile dysfunction in men, heart disease and stroke. Diabetes increases your risk of atherosclerosis and high blood pressure: diabetics have double the risk of cardiovascular disease than the general population and most die from heart disease. Overall, the death rate amongst people who have diabetes is around twice that in the general population – and this is mainly due to the increased risk of heart disease.

Lifestyle, diet and diabetes: Lifestyle factors are hugely important in the prevention and treatment of type 2 diabetes. A recent report from the World Health Organization called the modification of lifestyle 'the cornerstone' of both the prevention and treatment of diabetes. Keeping your weight down, getting plenty of exercise and eating healthily, with plenty of salads and vegetables, can halve your risk of developing type 2 diabetes.

Obesity is an important factor in type 2 diabetes; as with heart disease, especially when the extra weight is around the midriff. Losing weight improves the body's sensitivity to insulin and reduces your risk of developing type 2 diabetes. And, regardless of how fat you are, physical activity also brings down your risk.

Eating lots of saturated fatty acids (mostly found in animal fats) is linked to a higher risk of developing diabetes, while oils from plants – high in unsaturated fatty acids and polyunsaturated fatty acids – are associated with a decreased risk. Non-starch polysaccharides (unprocessed, or minimally processed, carbohydrates, sometimes loosely referred to as dietary fibre) also protect you against diabetes. These carbohydrates are found in wholegrain cereals, fruit and vegetables.

Much has been made of the glycaemic index (GI) of foods. GI is a measure of how rapidly carbohydrates are broken down into glucose and absorbed into the bloodstream, and eating low GI foods certainly reduces your risk of developing diabetes. However, GI has assumed almost cultish

status, becoming a bit of a trendy bandwagon, leapt on by the health and diet-conscious – and by those wanting to exploit them. Although the GI is usually a good guide to whether a food is 'healthy', it doesn't always work that way; for instance, high-fat foods can have a low GI. The content of non-starch polysaccharide in the food is more important. Sometimes the jargon and the endless lists of foods over-complicates what is really quite a simple message. Even doctors can get confused by the plethora of information about nutrition but, thankfully, a recent publication in *The Lancet* boiled the information down to some very simple advice. If you want to eat a healthy diet with plenty of non-starch polysaccharides and a low glycaemic index (no confusing numbers here):

→ Eat breakfast cereals made from oats, barley and bran
→ Eat wholegrain bread
→ Eat fewer potatoes
→ Eat plenty of fruit and vegetables (except potatoes)
→ Eat lots of salads with vinaigrette dressing

Pancreatitis

In developed countries, most cases of inflammation of the pancreas, pancreatitis, are caused by gallstones and alcohol. The precise sequence of events leading to pancreatitis isn't fully understood, although in pancreatitis brought on by gallstones, it seems to have something to do with the gallstones blocking the end of the pancreatic duct and the pancreas swelling up with its own secretions.

Chronic, or long-standing, pancreatitis is mostly down to alcohol. Alcohol seems to upset the delicate balance between the digestive enzymes that the pancreas produces, and the inhibitors of those enzymes. The pancreas makes proteases, the enzymes that break down proteins coming into the gut. But those enzymes also have the potential to break down the protein in your tissues. So, the pancreas has to suppress the action of its own enzymes, to inhibit them – or it's in danger of digesting itself. If the balance is disturbed, the pancreatic ducts get blocked and calcified. Calcium shows up on X-rays

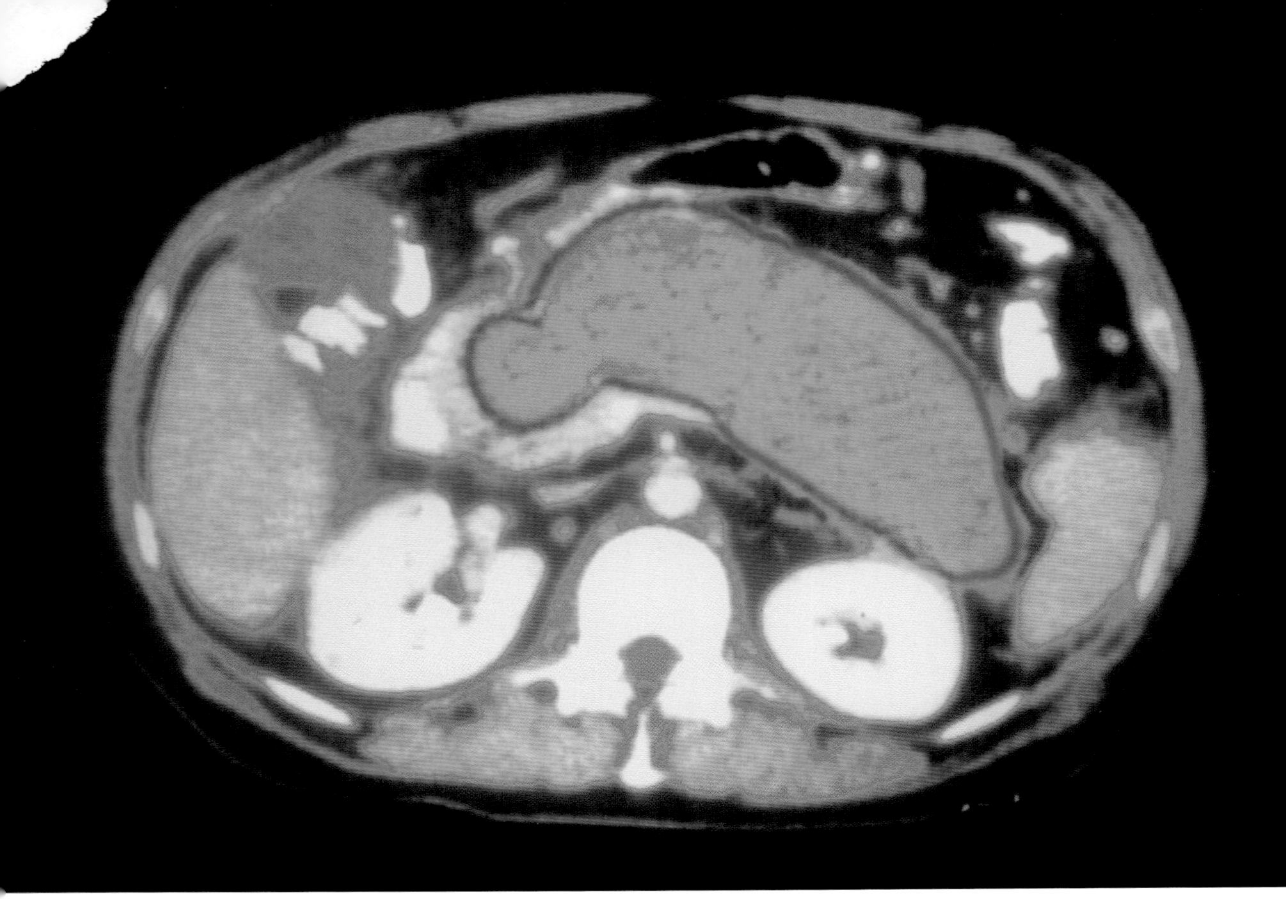

This coloured CT scan shows a slice through the abdomen, with a bloated, inflamed pancreas (in turquoise).

and CT scans and small white spots are seen along the pancreatic ducts in people with chronic pancreatitis.

Pancreatitis causes intense pain in the upper belly just under the sternum. People with pancreatitis often feel so ill that they don't eat and so lose huge amounts of weight. Both the digestive and endocrine functions of the pancreas are upset: without pancreatic digestive enzymes, the body can't digest and release the nutrients from food, leading to malnutrition, and without enough insulin, diabetes will develop. Chronic pancreatitis may also lead to pancreatic cancer. It seems that pancreatic cancer may follow on from chronic pancreatitis, but your risk of pancreatic cancer seems to be largely dependent on your genes (though it is doubled if you smoke). A cancerous growth in the head of the pancreas may block the bile duct, causing jaundice.

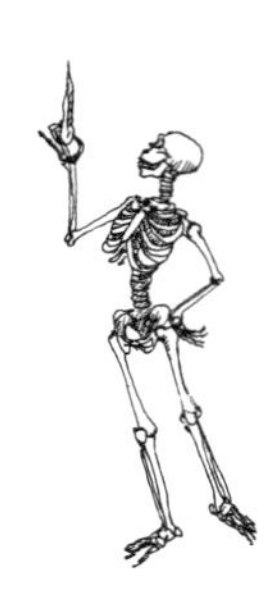

Five ways to keep your pancreas healthy:

- Keep your weight down; weight loss has been proven to reduce your risk of type 2 diabetes.

- Eat less fat and replace saturated fats with unsaturated fats; eat plenty of wholegrains, fruit and vegetables (except potatoes) and less refined sugar. This will reduce your risk of becoming insulin resistant and developing type 2 diabetes.

- Keep active – physical inactivity increases your risk of type 2 diabetes.

- Don't drink too much alcohol (a maximum of two units per day for women and three for men) and don't binge drink. This cuts your risk of developing alcohol-related pancreatitis.

- Don't smoke – it doubles your risk of developing pancreatic cancer.

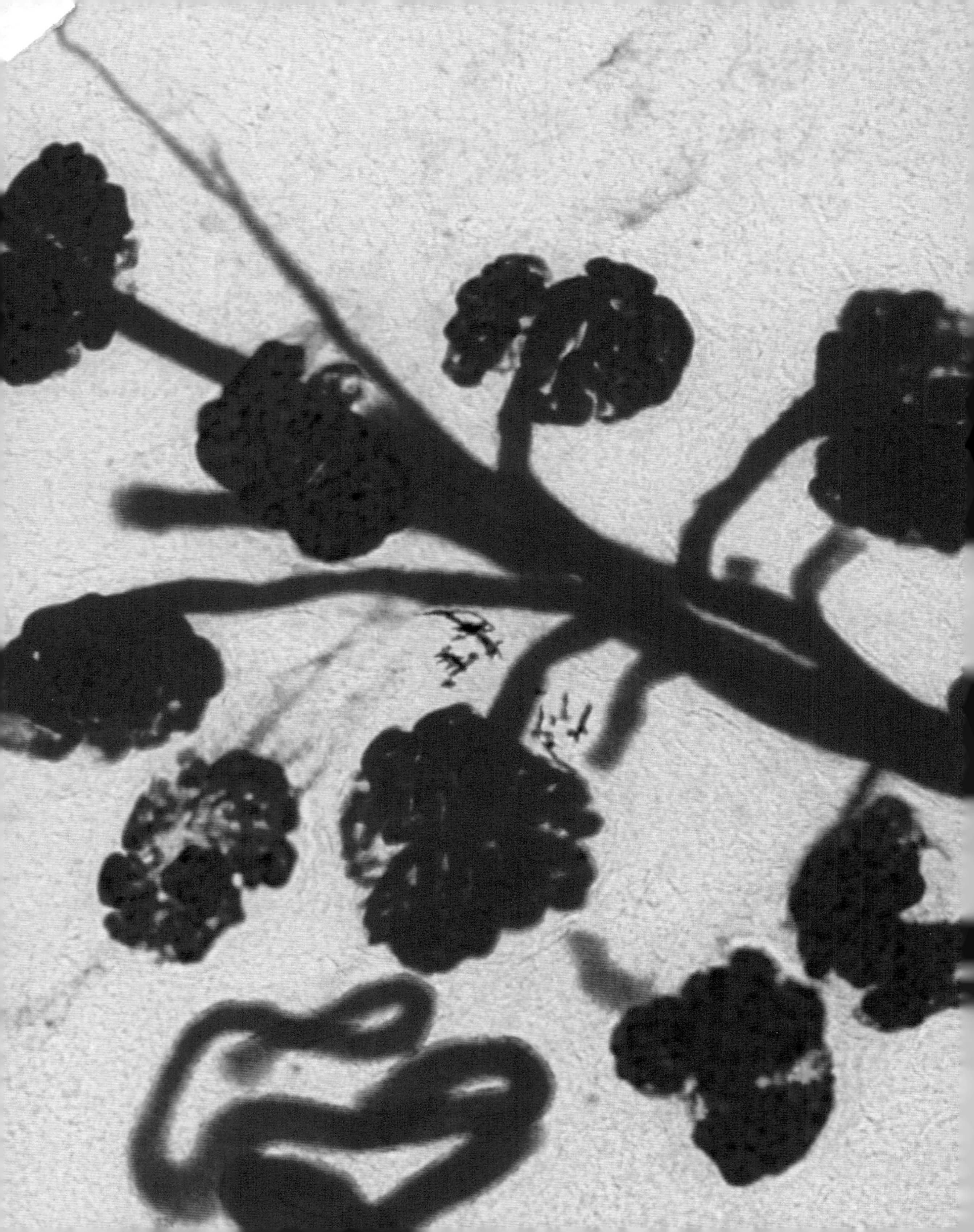

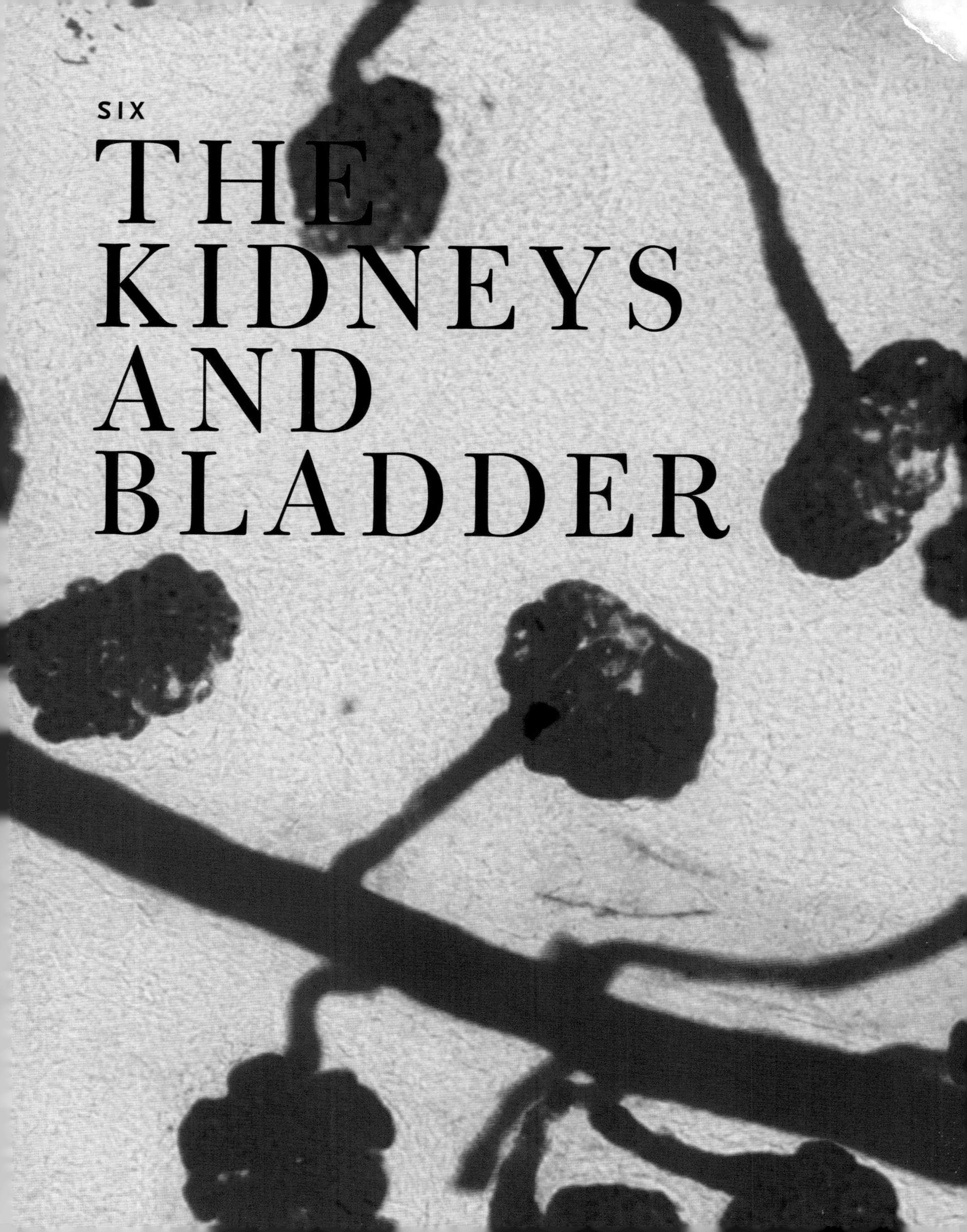

SIX

THE KIDNEYS AND BLADDER

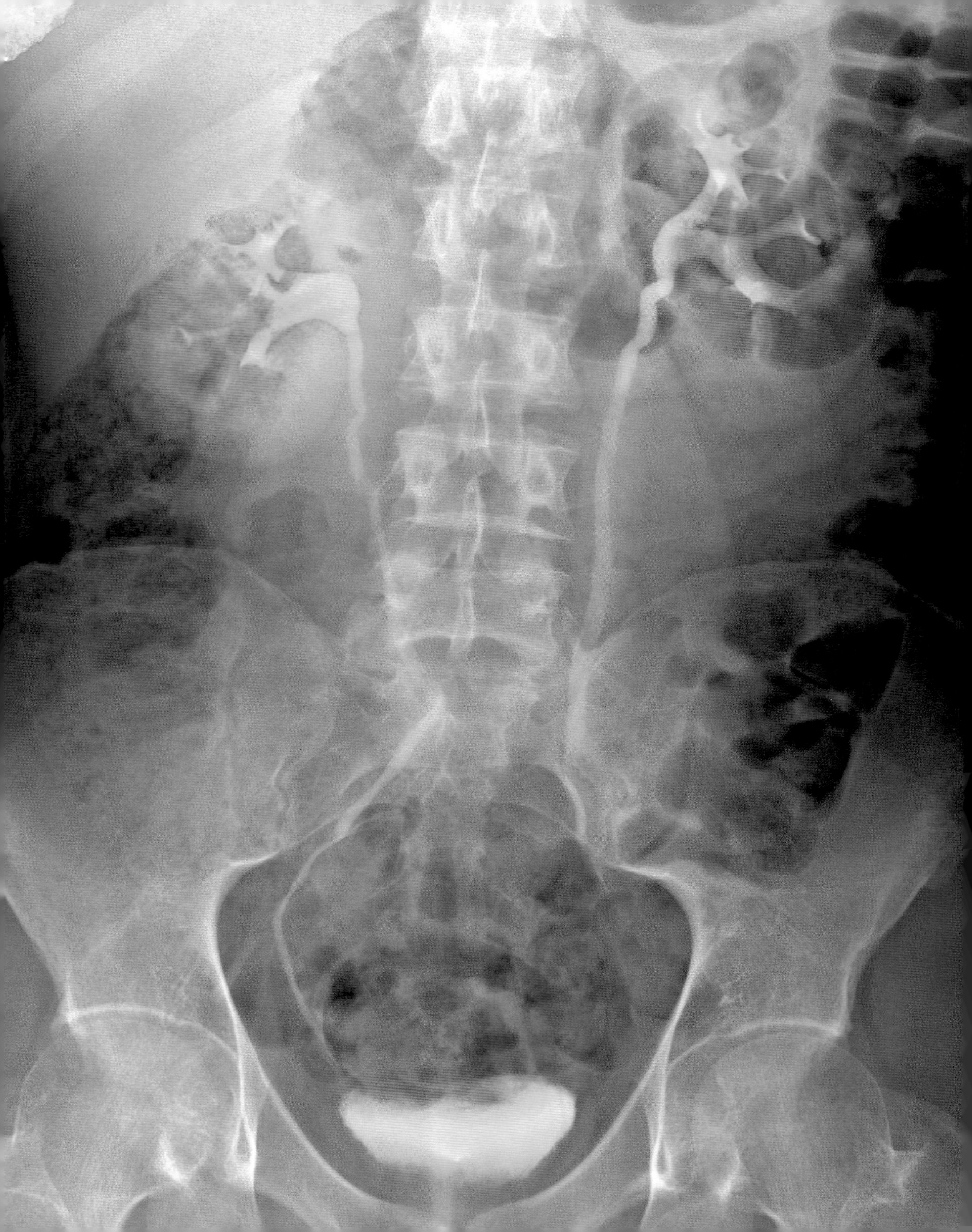

The kidneys do an amazing job of filtering your blood and making sure that it has exactly the right concentrations of charged ions to keep all the cells of your body in perfect working order. (Ions are atoms which have lost or gained an electron, making them electrically charged.) 'Electric' tissues like nerves and muscles rely on having the right level of ions inside and outside their cells to work properly, but the balance of ions (the electrolyte balance) in the blood and in the fluid around and inside cells is important to every cell in your body.

The kidneys regulate the volume of fluid in your body and play a very important role in excretion: getting rid of waste, especially the nitrogen-containing waste that your cells produce from the breakdown of proteins. This waste starts off as ammonia, which the liver makes less toxic by converting it into urea. Urea is carried in the blood to the kidneys, where it is excreted in urine. (Urea is a good source of nitrogen for plants – peeing on your compost heap really does make sense.) It's not surprising that if the kidneys fail to do their vital job properly, the whole body finds out very quickly.

Previous pages **Looking like miniature hydrangea blossoms or pom-poms, these balls of capillaries are the glomeruli, whose blood is filtered in the kidneys.**

Left **When an X-ray patient is injected with a special dye, the kidneys (orange, top of picture) glow as they filter the dye out into the urine. The urine passes down the thin tubes of the ureters into the bladder.**

Where are the kidneys and bladder?

The 'urinary system' consists of two kidneys with two tubes (ureters) connecting them to the bladder, and the final tube that leads to the outside world – the urethra.

The kidneys lie high up in the abdomen, right against its back wall and behind the intestines. They're so high up that the upper part of both kidneys is behind the last rib. This means that there's a lot of kidney below – and therefore unprotected by – the ribs but they have their own very special protection: a layer of thick perinephric (Greek for 'around the kidney') fat that is just about the last fat in the body to go during starvation. The bladder lies way below the kidneys, down in the pelvis, behind the pubic bones.

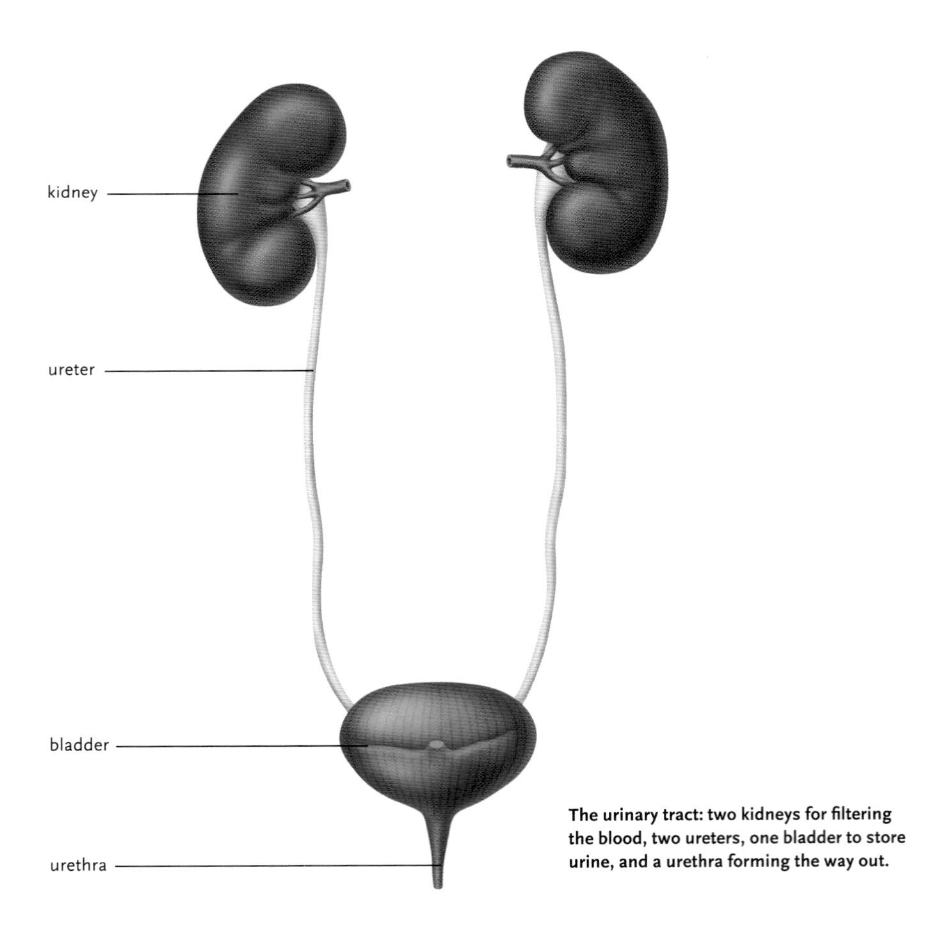

The urinary tract: two kidneys for filtering the blood, two ureters, one bladder to store urine, and a urethra forming the way out.

What do your kidneys look like?

The kidneys, not surprisingly, are kidney-shaped. Each one is about 10 centimetres long, 5 centimetres wide and 2.5 centimetres thick. They are dark reddish brown and have a smooth surface. If you were to open up a kidney, you would see the estuary where the urine drains through small tributaries into the large 'pelvis' (Latin for 'basin') of the kidney. This 'estuary' narrows as the urine enters the narrow tube that will carry it to the bladder. You will also see that the kidney is divided into a core or medulla, made up of pinkish pyramids, with a brownish outer shell, the cortex (which is Latin for 'rind'). What you can't see – unless you can look at a kidney under the microscope – is that the pyramids and cortex are a dense network of minute capillaries and tiny tubes, which filter the blood passing through the capillaries and remove waste and excess fluid. The tiny tubes are the nephrons; there are about a million of them in each kidney.

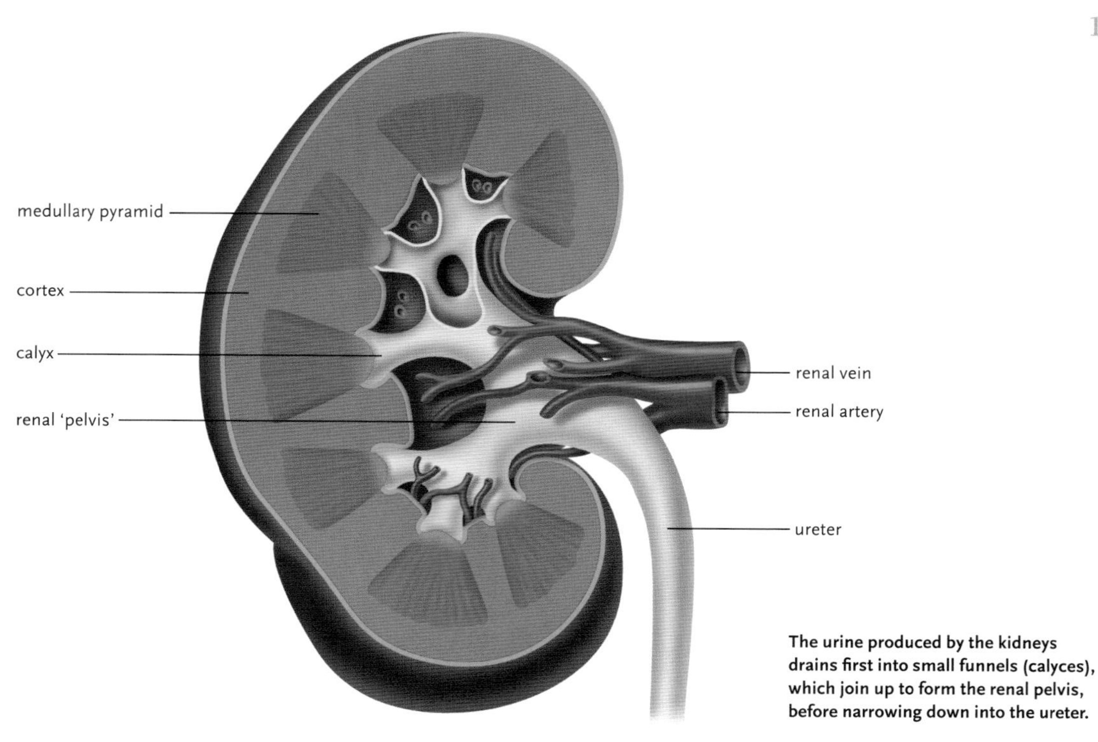

The urine produced by the kidneys drains first into small funnels (calyces), which join up to form the renal pelvis, before narrowing down into the ureter.

How the kidneys work

The blood flowing into your kidneys comes from the renal artery, which branches straight off the major artery of your body, the aorta. The renal artery plunges into the inner edge of the kidney, then branches and branches into ever smaller blood vessels until, eventually, it forms tiny capillary knots. The blood leaving these capillaries flows in tiny veins, which join up with other tiny veins and form the renal vein, which leaves the kidney and flows into the large vein, the inferior vena cava, that runs up the back of the abdomen.

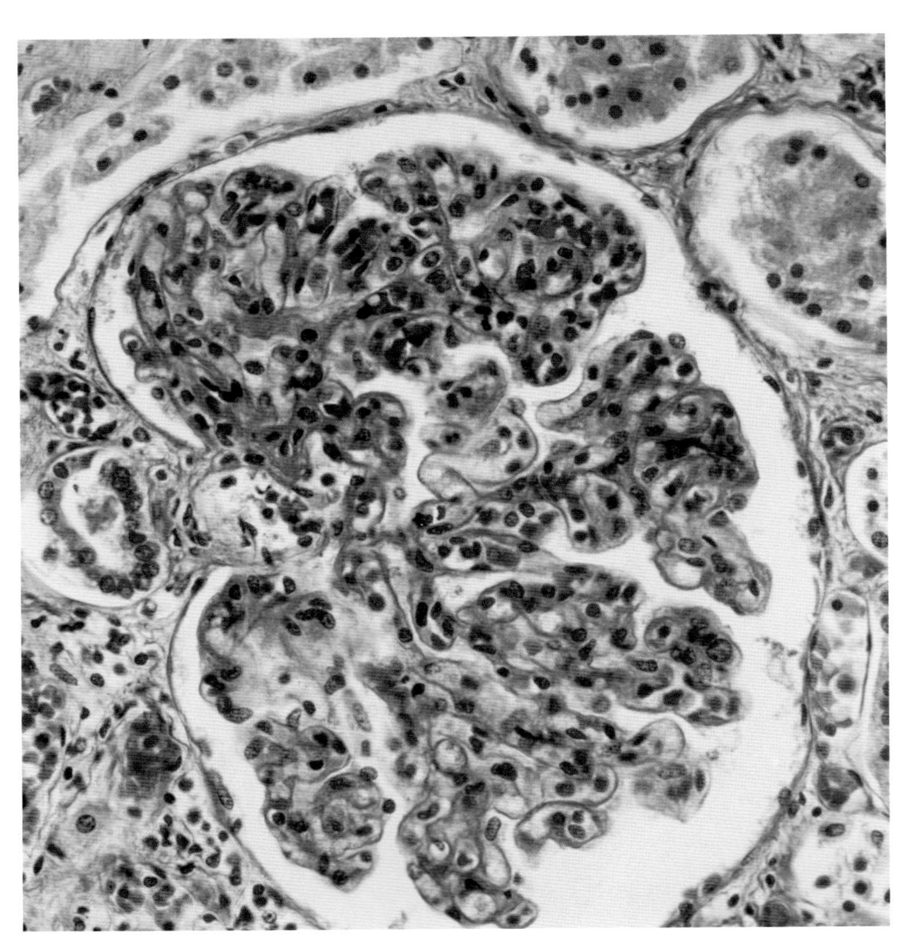

Filtering the blood: the tightly wound capillaries of the glomerulus (blue, with red blood cells inside) are embraced by Bowman's capsule (with the white space inside it).

Filtering the blood

Each tiny capillary knot sits in a cup-like structure, Bowman's capsule, which is the beginning of a nephron. This is where the important filtering job of the kidneys begins. The capillaries have leaky walls, which allow fluid to pour through from the blood. The fluid flows around the strange fern-like feet of the cells in the outer layer of Bowman's capsule, called podocytes (Greek for 'foot cells') and into the nephron. The kidneys filter an astounding 180 litres of fluid per day. This is much more fluid than you have in your body, so that can't be the end of the story.

As the fluid runs along the nephron, most of it is reabsorbed into the blood. This seems like a strange way to go about things but it means that the kidneys can remove waste from the blood even if it's there only in extremely low concentrations. The fine-tuning of the important charged ions –

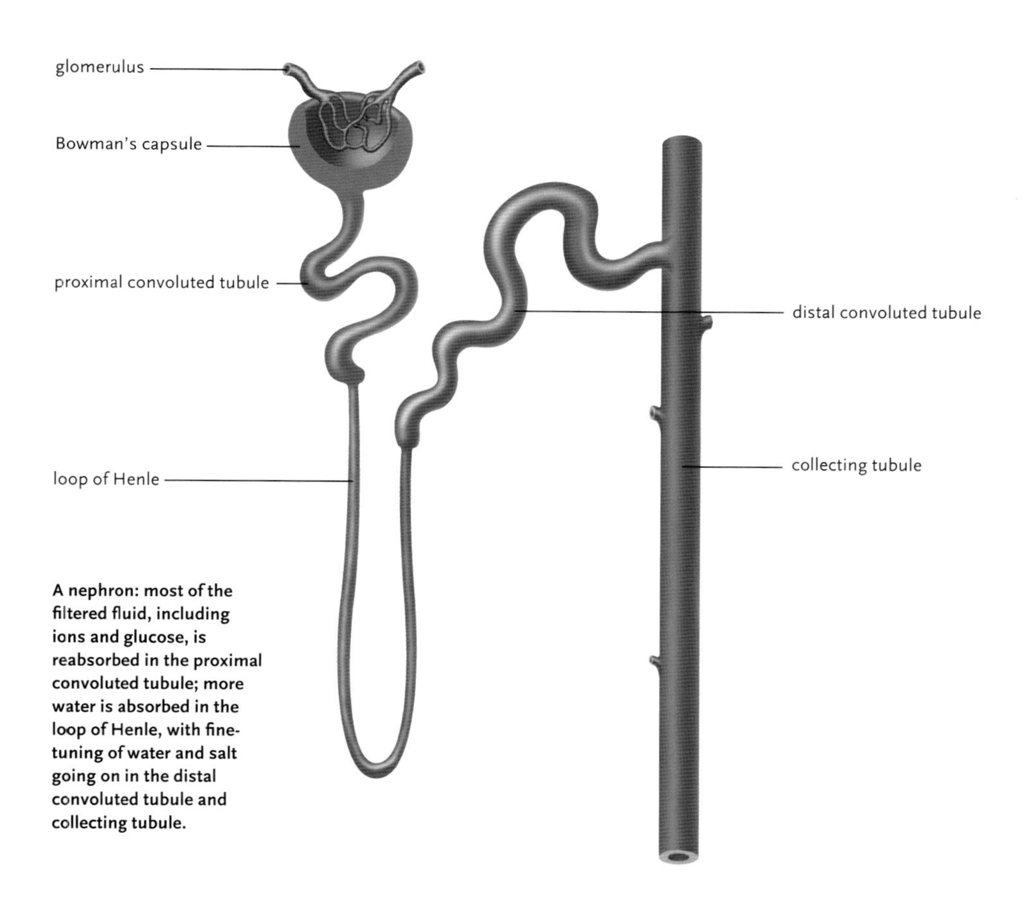

A nephron: most of the filtered fluid, including ions and glucose, is reabsorbed in the proximal convoluted tubule; more water is absorbed in the loop of Henle, with fine-tuning of water and salt going on in the distal convoluted tubule and collecting tubule.

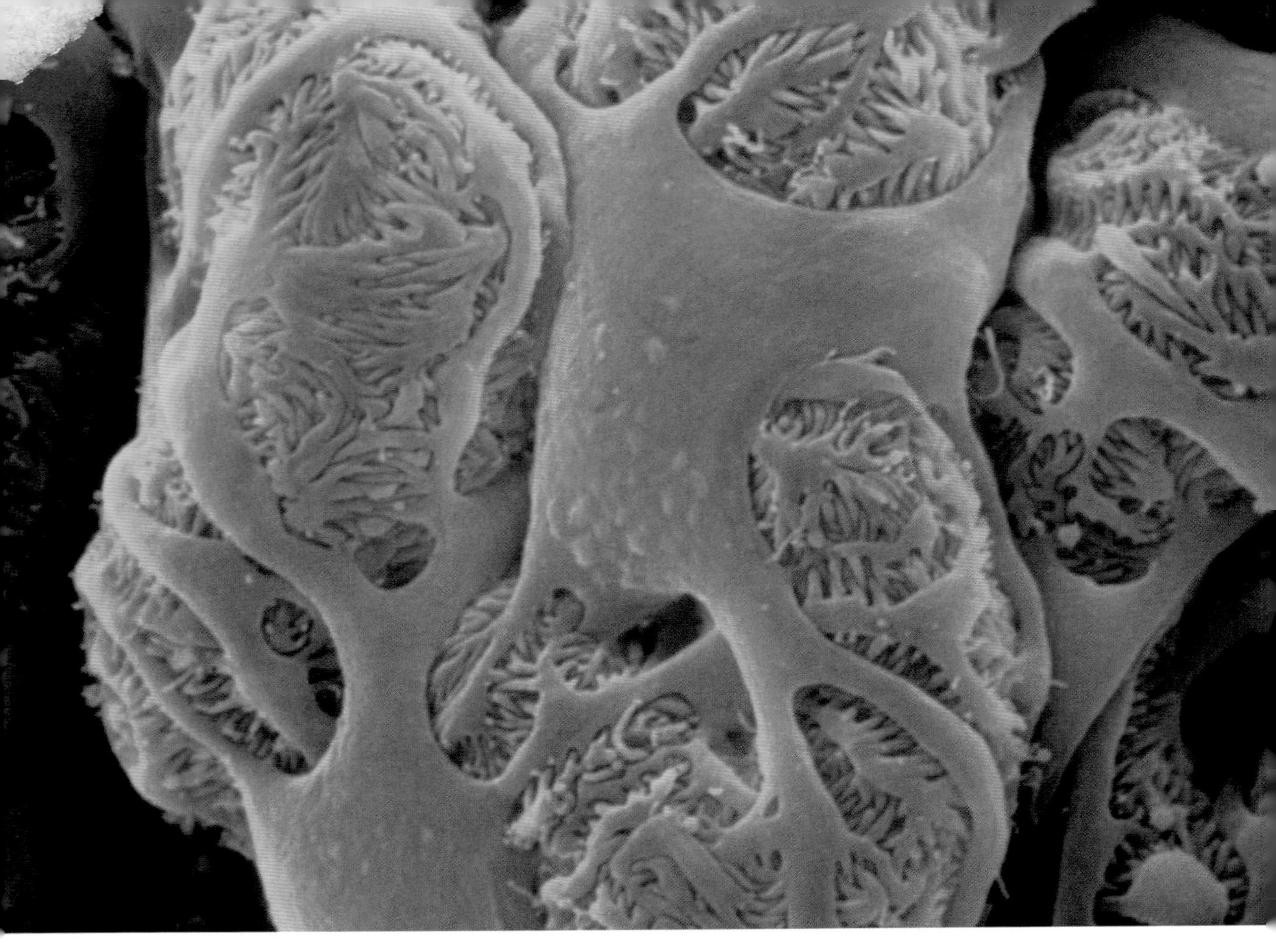

Fern-like foot cells: the fronds of the podocytes form a sieve through which fluid is filtered as it passes into the interior of Bowman's capsule, enlarged over 6,000 times.

sodium, potassium, calcium, magnesium and bicarbonate – goes on in the nephron, as it puts back exactly what the body needs into the blood. Glucose also gets into the nephron at Bowman's capsule but, normally, this glucose is reabsorbed into the blood leaving virtually none in the urine. In diabetes, high levels of glucose in the blood mean that the nephrons are overloaded; they can't reabsorb all the glucose, so some is excreted in urine.

When the filtered fluid reaches the end of the nephron, it flows into a collecting tubule, along with the fluid from lots of other nephrons. The collecting tubules run straight down the medullary pyramid to open on the papilla; when you look at the papilla under magnification, the tiny openings of the collecting tubules make it look like the top of a pepper pot.

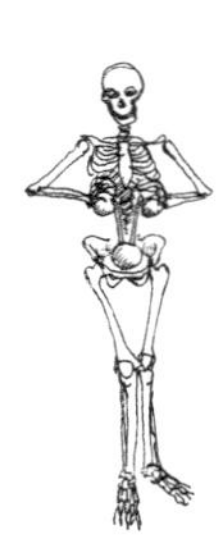

Pepper pot papilla: the tiny holes on the surface of the teat-like papilla are the openings of the collecting tubules, seen here magnified over 80 times.

By now, we call the fluid 'urine', and it flows into the calyces, which join together to form the wide renal pelvis, which collects all the urine from the kidney. The renal pelvis narrows down and forms the ureter (which comes from the Greek for 'water-maker'), which carries the urine to your bladder.

Your ureters are tubes of muscle; like the muscle wall of the intestine, they contract in waves to push the urine down into the bladder. This means that urine could still travel the 25 centimetres from the kidneys to the bladder even if you were standing on your head. The bladder is a muscular bag, which can hold about half a litre of urine and contracts, to release it, when you decide it's convenient. The urine leaves the bladder through the urethra, to reach the outside world.

Common kidney problems, and how to prevent them

Amongst the things that can go wrong with your kidneys, some are made more likely by various lifestyle and dietary factors. We'll take a look at kidney stones, cystitis and incontinence on our rounds.

Kidney stones

Kidney stones are a common problem affecting this organ; sometimes they are caused by disease or by genetic predisposition, but there are also dietary factors – factors that you can tackle to reduce your risk. Kidney stones are neither a new disease (there's evidence of them in ancient Egypt), nor an exclusively human one. Kidney stones are a mixture of inorganic crystals and proteins. Eight out of ten kidney stones are formed of calcium: they are white with a chalky texture. We have a calcium-based skeleton, which means a certain amount of calcium turnover has to go on in our bodies, and we need to get rid of some calcium in urine. As we're land animals, water is precious and we don't want to lose too much of it in urine, so our urine – including that dissolved calcium – is concentrated. As you raise the concentration of something that's dissolved, sooner or later you won't be able to dissolve any more (imagine putting spoonful after spoonful of sugar into a cup of tea). The stuff that can't dissolve will start clumping together (or precipitating) into solid particles, forming crystals. You can do things to make precipitation less likely, and one of the ways that the body puts the brakes on the formation of calcium crystals is to add citrate to the urine: citrate binds to calcium and keeps it soluble.

When a kidney stone leaves the kidney and starts to get pushed down the ureter, it can cause pain – which can be excruciating. It is described as 'colicky', which means that it comes and goes in waves, as the contractions in the ureter's muscular walls try to push the stone along. The pain tends to be high up in the side and spreads down to the groin. Occasionally, stones stick in the ureter and urine builds up behind it. This can eventually cause kidney failure. Some stones are so large that they never make it out of the kidney:

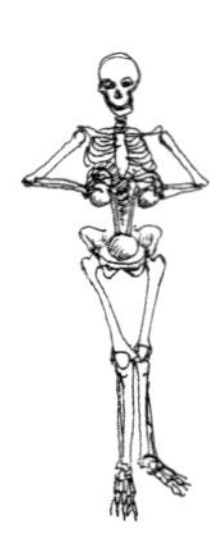

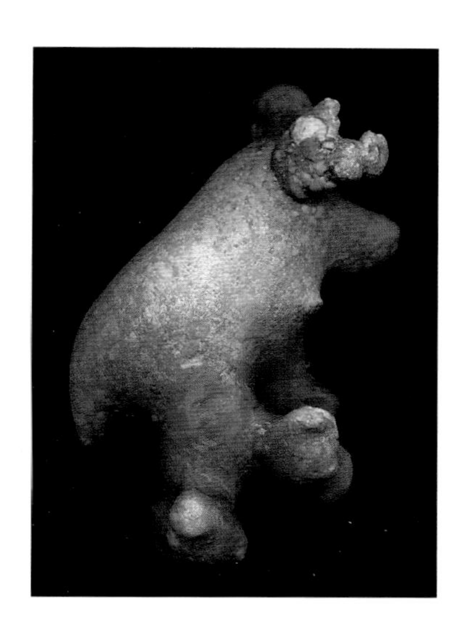

This staghorn-shaped kidney stone completely filled the calyces and renal pelvis of the kidney it formed in.

they get stuck in the renal pelvis and grow into the calyces, forming a large, branched stone, known in the trade as a 'staghorn calculus'.

Kidney stones are about four times as common in men than women – up to the age of 60, when the rate in men reduces and women catch up. They often occur in people who have otherwise been perfectly healthy, although they are also associated with diseases such as bladder and kidney infections. The risk of developing kidney stones at some time in your life is about one in ten and if you've had them once, you're more likely to get them again.

Kidney stones can be caused by obstruction of the free flow of urine, something which can be associated with enlargement of the prostate gland, a common problem in elderly men. Infections of the urinary tract are another cause: bacteria produce carbonates, which form stones. However, the main cause of kidney stones is a high level of calcium in the blood (often combined with a low level of citrate in urine). High calcium levels may be linked to underlying problems in the kidneys, problems with the bones (if calcium is being lost from bone into blood) or problems in the gut (if too much calcium is being absorbed).

Kidney stones are more likely to form if you are dehydrated for long periods. The kidneys work hard to make sure that the composition of the blood is correct; if your whole body is dehydrated, the amount of fluid your kidneys extract is reduced and your urine becomes very concentrated. You have to maintain the balance between water absorption and water loss: you need to drink more if you're losing lots of water from your gut, for example if you're sweating heavily or have diarrhoea. Kidney stones are more likely to precipitate from highly concentrated urine – so keep drinking! The NHS guidelines suggest trying to drink 1.5 to 2 litres of water per day, but the most important thing is to listen to your body and respond to its thirst. People who have kidney stones are generally advised to drink even more (two to three litres per day) and, contrary to popular opinion, it seems that this doesn't have to be plain water: orange juice, coffee and alcohol are all fine.

High levels of salt and protein in your diet can also make kidney stones more likely. Cutting down on salt means that less sodium goes into your urine; and because sodium and calcium go hand-in-hand, this reduces the levels of calcium. (It's possible that this effect of salt restriction, that is, reducing the amount of calcium lost in urine, may also reduce the risk of osteoporosis or weakened bones.) Some kidney stones are formed from uric acid, which is made when proteins are broken down, so cutting the levels of protein in your diet can reduce the risk of forming this type of stone. Protein restriction also raises citrate levels in urine, which helps stop calcium crystals forming. Eating lots of fruit and vegetables helps, as it makes your urine less acidic. Calcium restriction sounds like an obvious thing to try but, unfortunately, cutting down on calcium isn't a good idea and can cause an increase in other substances (such as oxalates) that also cause kidney stones.

At the end of the day, if you're healthy and you've never had kidney stones, you shouldn't really worry about these types of dietary changes. If you need to change your diet because you're at risk of kidney stones or have had them before, your doctor can advise you on how to do so.

Cystitis

Cystitis, inflammation of the lining of the bladder, is much more common in women than in men because the tube to the outside world, the urethra, is so much shorter in women. (Female urethras are about 4 centimetres long, male urethras around 20 centimetres – the male urethra has to travel the length of the penis.) The infection is usually caused by bacteria from faeces, such as *Escherichia coli*, that have managed to make their way forwards towards the opening of the urethra (which lies between the clitoris and the vagina).

The urethra has defence mechanisms – including mucus to trap bacteria, continual shedding of cells from its lining, immune protection and,

of course, a regular wash-out with urine – but sometimes the bacteria just get the better of these defences. Infections of the urinary tract usually lead to a need to pee often, and with a painful or burning sensation when the urine is passed. Sometimes there is pain in the suprapubic area, the very lowest part of your abdomen, just above the pubic bones, which is where the bladder lies. There may also be blood in the urine or an unpleasant smell to it.

If you're female, to avoid cystitis or, at least, to reduce your risk as much as possible, you need to look after the area where your urethra and vagina both open: the 'vestibule', between the labia minora. Keeping a clean vestibule (which is Latin for 'hallway') lowers your risk of cystitis, but be wary of using perfumed bubble bath and talcum powder in that area. You should remember to wipe from front to back after going to the toilet, to avoid bringing bacteria from the anus forwards into the vestibule. Using a lubricant during sex helps to prevent damage which might let infection

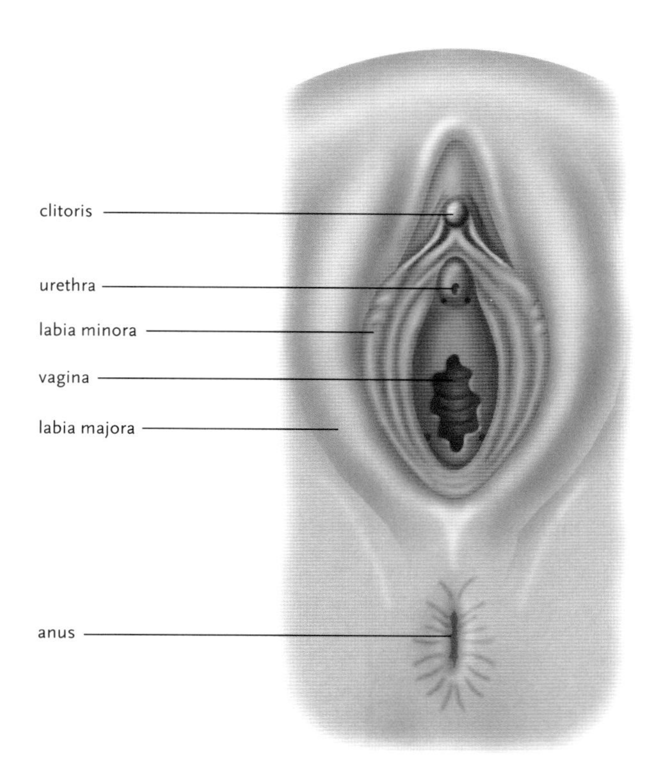

Between the legs: the labia minora enclose the vestibule, into which the urethra and vagina open. The anus is not far behind.

in more easily, and emptying your bladder as soon as possible after sex washes out any unwelcome visitors in your urethra.

The complications of a urinary tract infection can be very serious: the infection might not stop at the bladder and the bacteria may crawl up the ureters into the kidneys. Kidney infections are extremely painful and can also cause a fever. Small abscesses form, full of pus; the kidney becomes inflamed and struggles to maintain your electrolyte balance and excrete waste. Sometimes – especially if there are other problems such as kidney stones or diabetes – infections can cause lasting damage to the kidney.

Cranberry juice has acquired something of the status of a mythical miracle cure when it comes to cystitis but is there any evidence to suggest that this particular fruit juice, unlikely as it sounds, might actually work? Well, for once – yes!

Cranberry juice contains fructose (like every other fruit juice) and an unusual, large molecule (also found in blueberry juice). These both seem to stop bacteria like *E. coli* latching on to the lining of the bladder. The unusual large molecule was a bit of a mystery for a while but has now been identified as a type of flavonoid or plant pigment (for those that really want to know, a trimer of proanthocyanidin) and it is an antioxidant as well.

An amazing trial, using a placebo (a neutral substance) coloured and flavoured to taste just like cranberry juice, was carried out in an elderly people's home in Boston. The elderly female residents drank 300 millilitres per day of either the real stuff or the placebo (but didn't know which) for six months. The results showed that not only did drinking cranberry juice help to prevent cystitis in the first place, it also seemed to cure existing infections. This study was carried out among elderly women, with an average age of 79; the experiment hasn't been tried in younger women. But if you do get recurrent cystitis, drinking cranberry juice certainly won't do you any harm and it seems that it could do you a lot of good. But I'm not advocating chucking out the conventional medicines. Do go and see your GP if you think you've got cystitis and do take any antibiotics you're prescribed. And drink plenty of fluids to wash those bacteria out of your system.

Incontinence

To fully understand incontinence (when things go wrong with the bladder), you need to get a grip on how continence works by understanding a bit about normal bladder anatomy and physiology. The bladder is a bag made of smooth muscle, by nature 'automatic', and supplied with autonomic nerves. Stretch-receptors in its walls start up a reflex in these nerves that make it contract. This is how our bladders work when we are babies: they fill up . . . they empty . . . they fill up . . . they empty, without conscious thought. Potty training is all about learning to overcome this primitive reflex. With a little mental effort, the automatic emptying of the bladder can be overcome and releasing urine starts to be something we have control over. After a while, we don't even notice the mental effort of suppressing the reflex and peeing becomes a voluntary act: we can do it when it's convenient instead of when the bladder 'wants to'.

The bladder sits on a sheet of skeletal (voluntary) muscle known as the pelvic floor or pelvic diaphragm (anatomists call it the levator ani – the 'lifter of the anus'; try tensing it – it does!) The urethra goes though the pelvic diaphragm, so when this sheet of muscle contracts, it acts as a sphincter around the urethra. (You can test this sphincter action for yourself; contract your pelvic diaphragm so that it squeezes the urethra closed and you should be able to stop peeing in mid-flow.) In women, the fascia around the bladder and its neighbouring organs, in particular the vagina, helps to support the bladder and the urethra – keeping them in the right place – which is very important for continence.

Evolution progresses by small increments, minute alterations to the plan of animals' (including human) bodies, and it doesn't anticipate problems that might arise some time in the future as each change is made. Humans evolved to walk on two legs, which works better with a narrow pelvis – but then evolved to have large brains. Our babies have relatively large heads, and those heads need to fit through the female pelvis. This means that there's a bit of a design compromise in the female pelvis: walking on two legs works well with a narrow pelvis, but the pelvis needs to be wide enough to form the

Following pages **Urine has been made to show up on an X-ray by injecting the patient with a radio-opaque dye that is filtered by the kidneys. The X-ray on the left shows the bladder filling; on the right, after urination, the bladder is empty.**

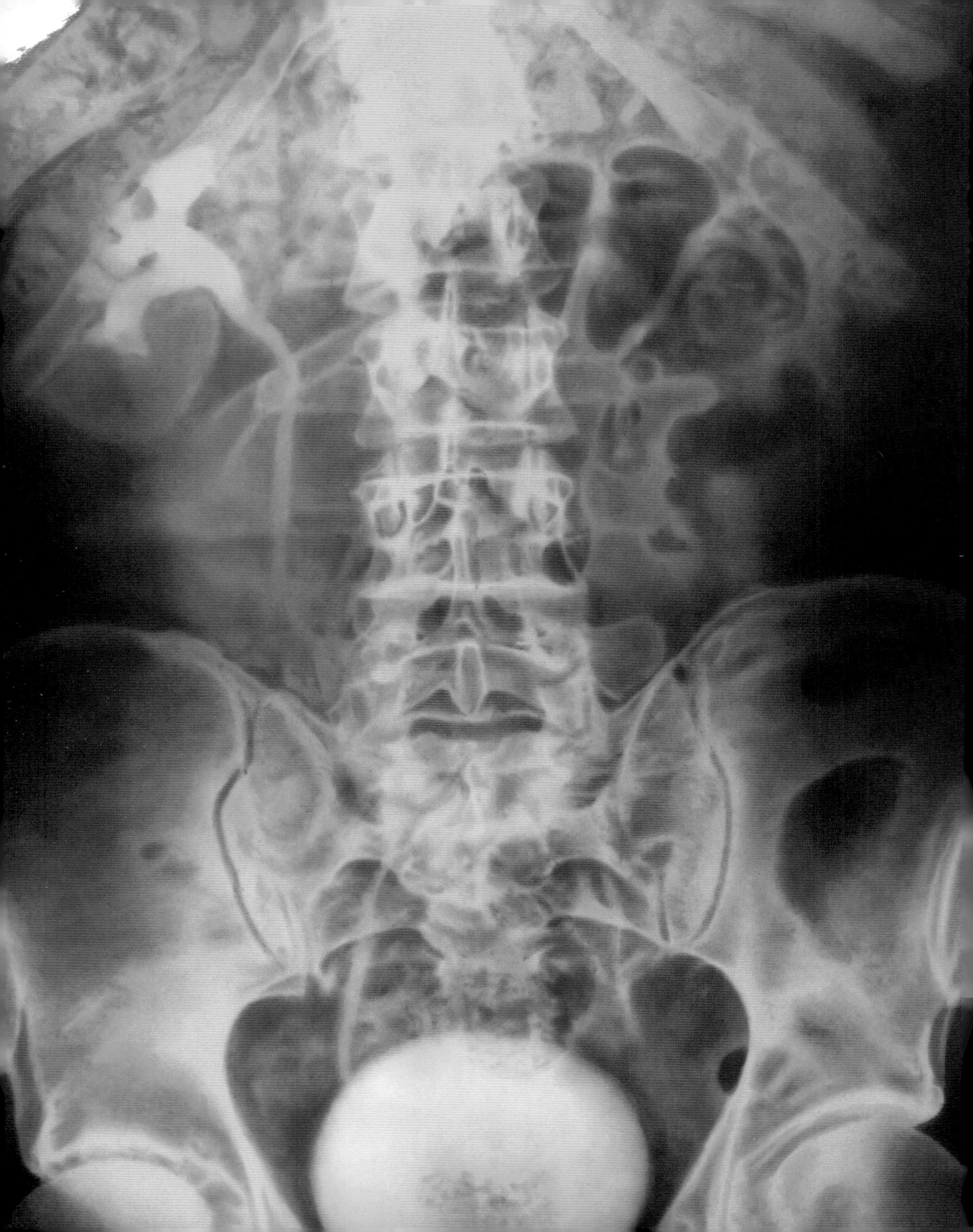

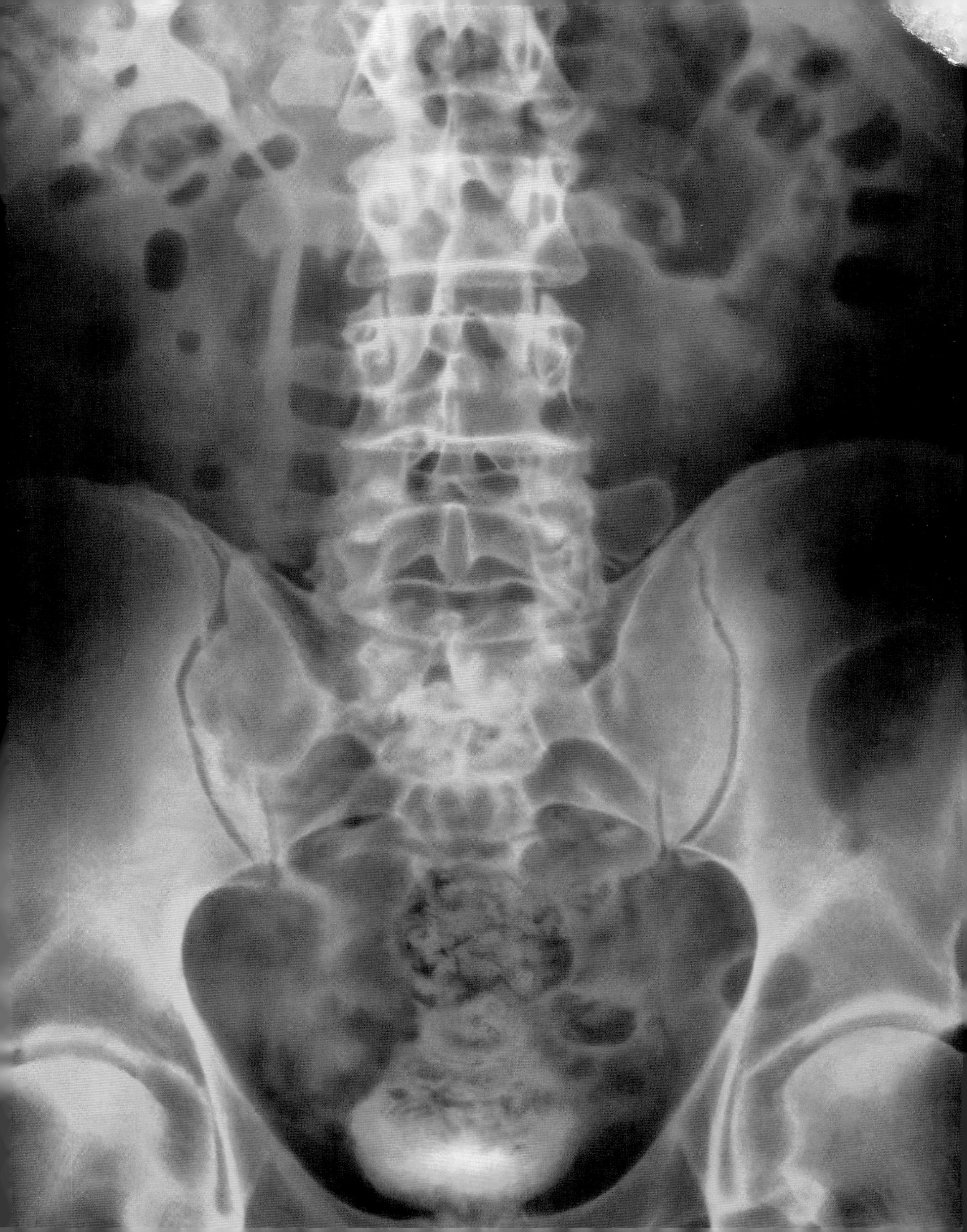

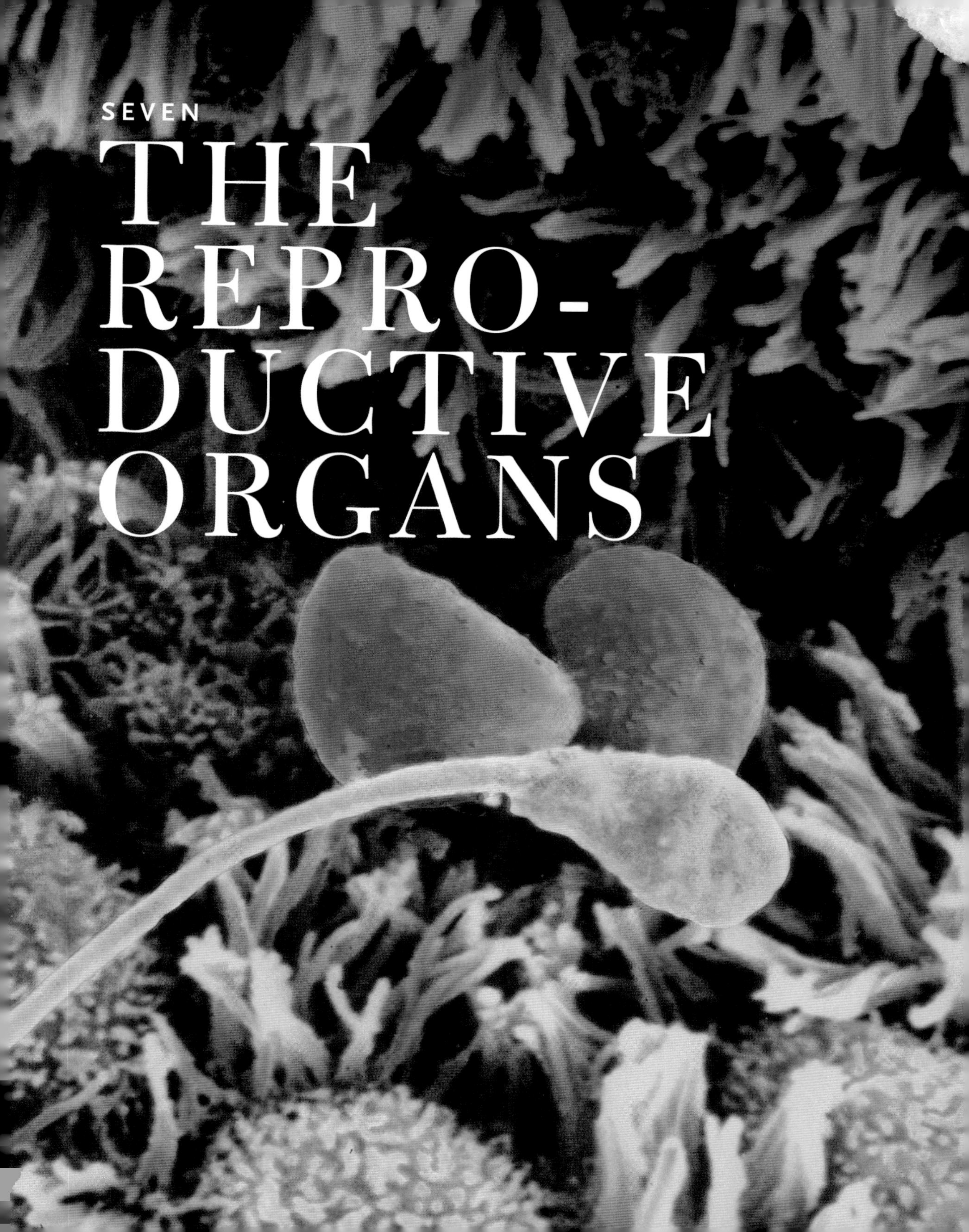

SEVEN

THE REPRODUCTIVE ORGANS

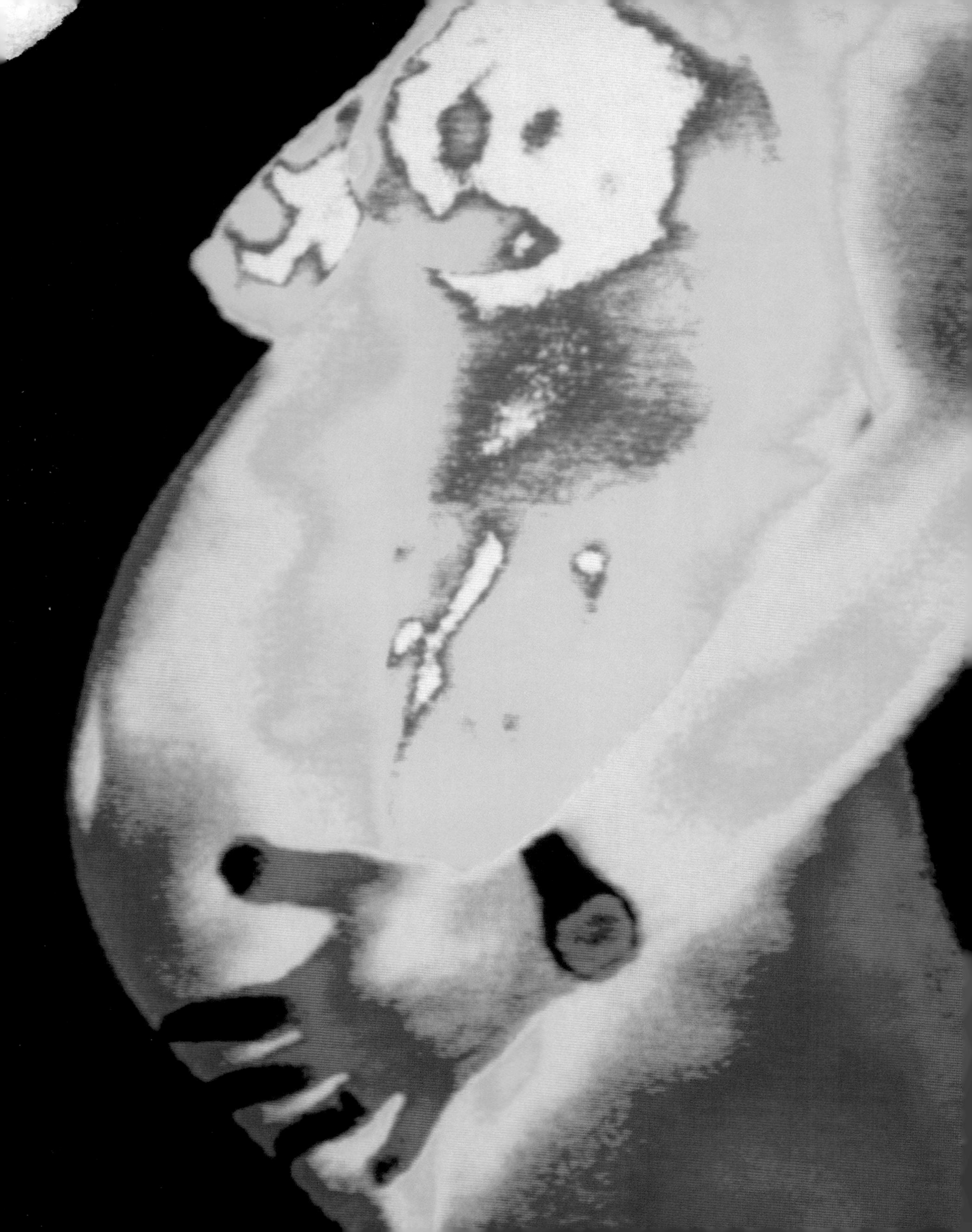

So here's the exciting chapter (did you turn to this one first?) Of all the organs in the human body, these are the ones that intrigue people the most. Sex is a wonderful part of being human – and the organs that enable it to happen are quite wonderful as well.

All the other organs are similar in both sexes. A man's liver, heart or lungs may be a bit larger than a woman's, but apart from that, they're pretty much the same. But when you get to the reproductive organs, there's a world of difference.

Both sexes boast a set of gonads and a set of tubes to go with them, but the arrangement is very different in men and women. The female gonads, the ovaries, are inside, tucked away in the pelvis. The male gonads, the testes, hang around on the outside in a specially crafted bag. Both ovaries and testes make cells – ova (eggs) and sperm – that have the miraculous potential to combine with each other to create an embryo. The gonads also make hormones: oestrogen and progesterone from the ovary; testosterone from the testis.

Previous pages **Incredible journey: this sperm is magnified over 10,000 times; it has swum all the way from the cervix, through the uterus, into the oviduct (which is carpeted with cilia – in pink).**

Left **Positively glowing: a thermogram of a pregnant woman.**

Where are the reproductive organs?

The male and female bits (the genitalia, as they're known in the medical profession, which sounds to me like a shortening of 'genital paraphernalia') are very different and kept in very different places. The male bits almost all hang about on the outside (sperm like to be cool, so the testes live outside the body, in the scrotum), whereas most of the female bits are hidden away inside the pelvis (ova seem to be perfectly happy at body temperature).

The male tubes are quite simple at the start; they're basic tubes that carry the sperm from the testes to the base of the penis, just beneath the bladder in the pelvis. Then there's a complicated bit of equipment, the penis, designed to get these sperm inside the female. The female tubes, the oviducts or Fallopian tubes, collect eggs from the ovaries and transport them to the uterus (where the foetus grows); then there's another tube – the vagina – connecting the uterus with the outside world. The vagina is both the way the sperm gets in and the way the baby gets out.

It seems quite amazing that these different-looking and differently positioned bits could have come from the same components in the embryo – but they did. Seven weeks after conception, the embryo has a beating heart, arms, legs, brain and spinal cord and a completely ambiguous set of genitals – a long, sausage-shaped gonad and some tubes on the back wall of the abdomen – which could go either way. After week seven, the genes that control sexual development kick in: in a male fetus, the gonads become testes that migrate out of the abdomen into the enlarging scrotum, while each tube turns into the vas (or ductus) deferens; in a female fetus, the gonads become the ovaries and migrate a little way down into the pelvis, while the tubes partly fuse together to make the vagina, uterus and oviducts.

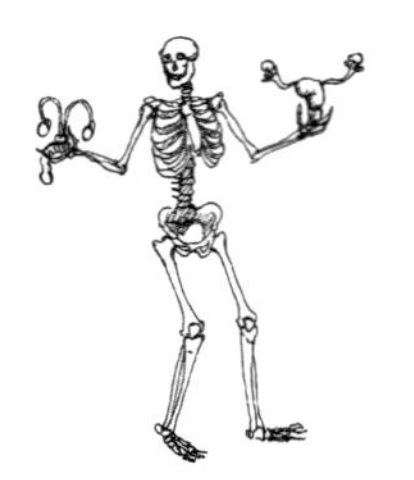

In a ten-week embryo, the genitalia (shown over 200 times actual size) have not yet been 'told' which sex they are; if male, this genital tubercle will become a penis; if female, a clitoris.

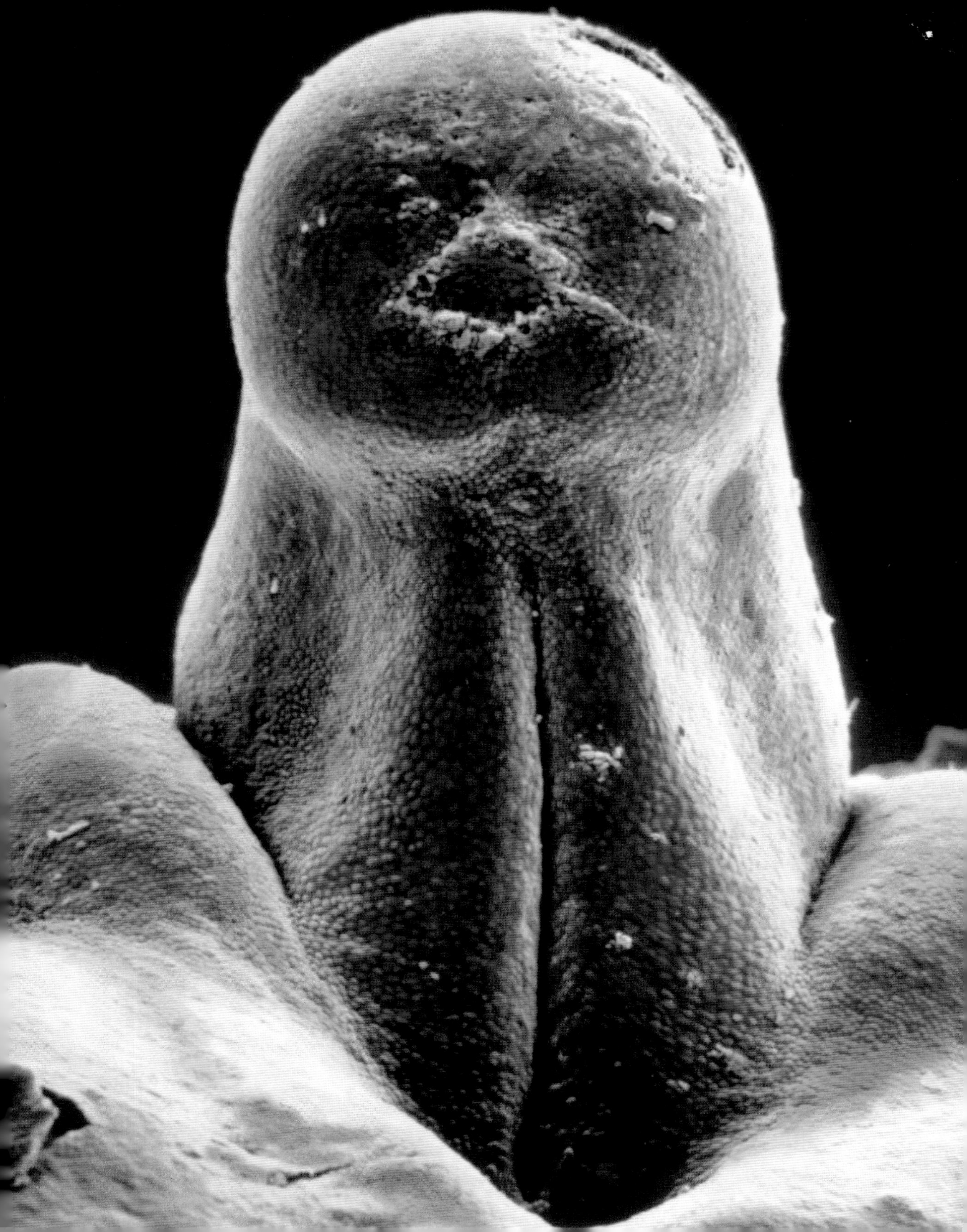

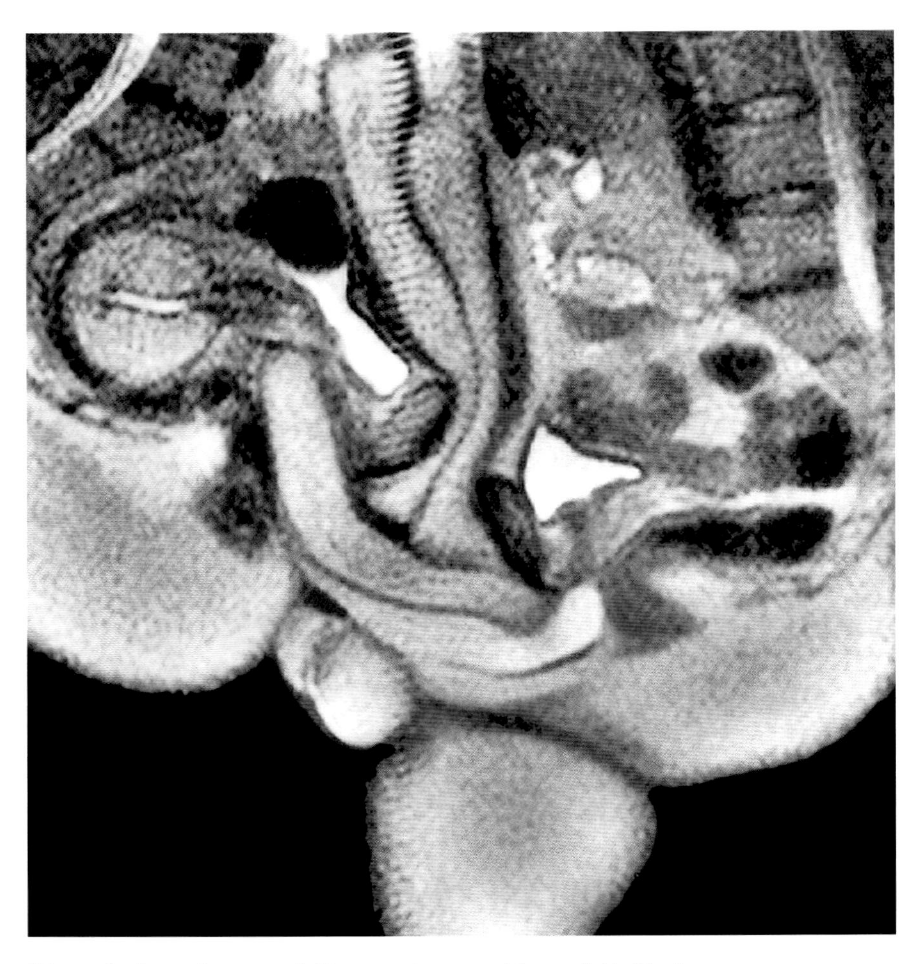

This amazing image shows a couple (the woman in green and the man in blue) having sex . . . in an MRI machine. The penis is practically bent into a boomerang shape.

testes contracts, pushing the sperm out of their holding station in the epididymis (Greek for 'upon the testis'), the elongated lump running down the back of each testis, and into the vas deferens (Latin for 'the carrying away vessel'). The vas contracts in waves, pumping the sperm along until they reach the upper part of the urethra, enclosed within the prostate gland, just under the bladder. As the sperm pass along, various glands secrete fluids, which contain nutrients for the sperm and bicarbonate to combat the acidity of the vagina. The seminal vesicles behind the bladder, the prostate gland and the pea-sized bulbourethral glands (so-called because they empty their

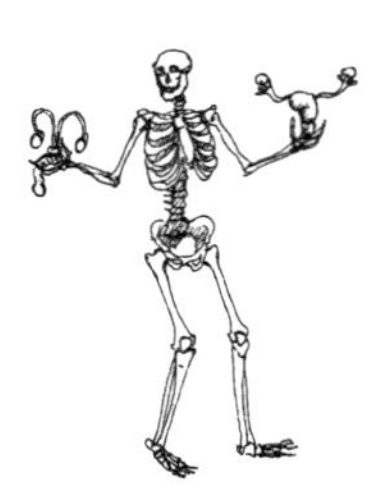

secretions into the urethra just below the prostate and perineal membrane, where the urethra lies within the first part, or 'bulb', of the penis) all add a contribution of fluid to the sperm. More waves of muscle contraction along the urethra, aided by rhythmic contractions of the perineal muscles, pump the sperm along the last leg of its journey, to the tip of the penis and out.

The female side of things is by no means passive. There is a crescendo of reflex muscular activity, until the vagina, pelvic floor and perineal muscles produce waves of contraction. For both the male and the female, the climax to a complicated series of reflexes is accompanied by a feeling of intense pleasure: orgasm.

Fertilization

If everything is in good working order, about two millilitres of semen, containing around two million sperm, is delivered to the upper part of the woman's vagina. The sperm still have a long way to go and now they have to travel under their own steam. Tails thrashing furiously, they swim up through the cervix (the neck of the uterus), through the uterus to its upper corners, through the narrow opening of one of the oviducts and into a wider area, the ampulla. Not all the sperm get all the way – the vagina, cervix and uterus are littered with the microscopic corpses of sperm that haven't made it.

For those that have, the race is on. If it's the right time in the woman's ovarian cycle, if she's fertile, if she's not taking the contraceptive pill and if the sperm are in the correct oviduct (remember there are two – and usually only one ovum) – then they've got a chance of fertilizing an egg. The 'best man wins' when the first sperm reaches the ovum and successfully penetrates its gel-like outer membrane. This initiates a change in the membrane that makes it impenetrable to more sperm. This is very important, because additional sets of male chromosomes entering the ovum are not only unnecessary but will produce genetic defects. Once the chromosomes from the sperm have floated into the egg, they partner up with the female chromosomes, like pairs at a dance, ready to start replicating in preparation for the first cell division. Within 24 hours, the fertilized egg has divided into two new cells – the beginning of an embryo.

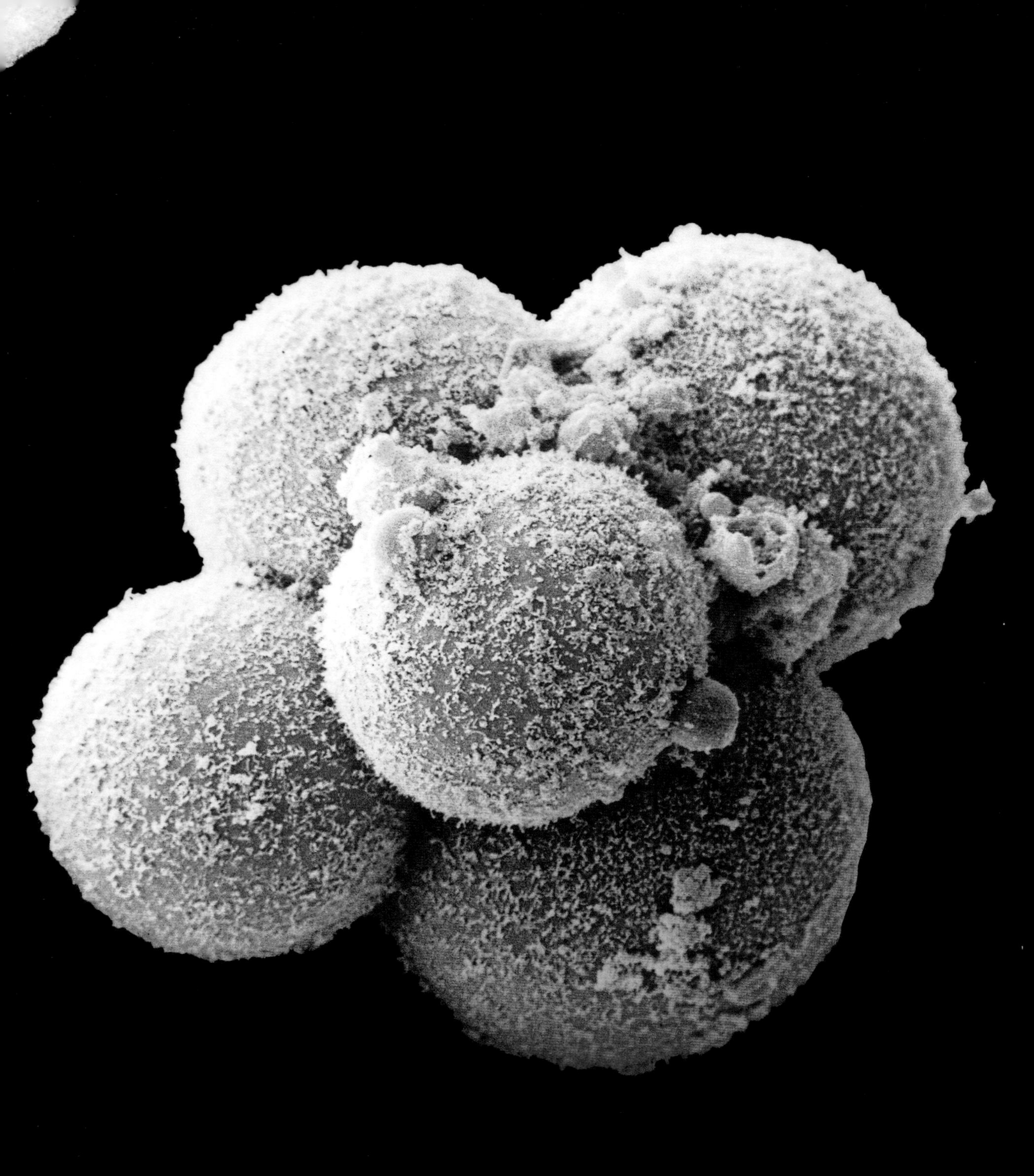

Common reproductive organ problems, and how to prevent them

With all those tubes and complicated reflexes, it's not surprising that things can go wrong. It's something people can be quite shy of talking about – when a man has erectile dysfunction, when a woman doesn't get orgasms or it hurts to have sex, it's very easy to think you're the only one in the world with this problem. But look at the statistics: over the age of 40, at least four in ten men experience problems 'getting it up'; nearly half of women have period pain and some experience periods so painful that they need to take strong painkillers and may not be able to get on with their normal lives for several days a month; many women (the numbers are hard to judge but it may be as many as one in five) find having sex painful, which can feel like a terrible thing to admit about something which should be so pleasurable. And one in ten couples have low fertility. There are lots of causes for this, including low sperm count, blocked oviducts or genetic incompatibility between the two. Your lifestyle affects many of these problems; I shall take a closer look at infections, period pains, erectile dysfunction and cancers.

Infections

Sexually transmitted diseases (STD) are on the increase, particularly in 15- to 25-year-olds. Chlamydia and gonorrhoea, both bacterial infections, have seen a huge rise and viral infections like herpes and HPV – not forgetting HIV – also increased between the late 1990s and the early 2000s. If you're diagnosed and treated for an STD, it's really important that your partner (or anyone with whom you have had recent sexual contact) gets treated as well or there's a risk you could be re-infected. If your partner's being treated, you should be as well. You can have these infections without having any symptoms, so it's best to be safe.

Chlamydia: *Chlamydia trachomatis* is a very common sexually transmitted bacterial infection, the most common bacterial STD in developed countries. The incidence in the UK has risen rapidly. Between 1997 and 2002, the

Shown a few days after fertilization, this ball of cells will carry on dividing and growing to make a baby.

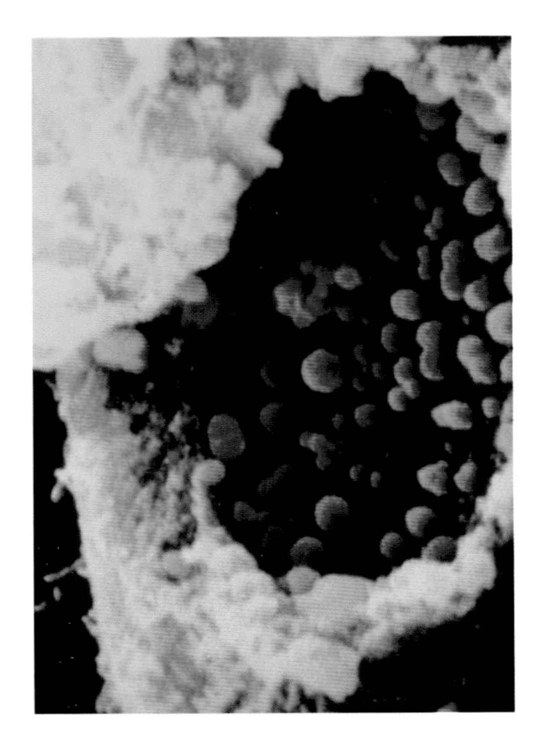

The enemy within: an infected cell broken open, and positively seething with chlamydia bacteria (the tiny red spheres).

number of new cases of chlamydia more than doubled. It's estimated that about one in 20 sexually active women in the UK are infected. Sometimes there are no symptoms but women with a chlamydia infection may experience vaginal discharge, pain on sex and a burning feeling when passing urine. Apart from these unpleasant initial symptoms, chlamydia can also damage the oviducts, leading to fertility problems and increased risk of ectopic pregnancies. (Ectopic pregnancy means that the fertilized ovum lodges somewhere other than the uterus, such as in the oviduct; ectopic pregnancies carry a great risk of rupture and internal bleeding.) Chlamydia infection in men may cause a discharge of fluid from the urethra and pain on passing urine. If left untreated, the infection can track back along the vas deferens to infect the epididymis or testes. Chlamydia can be treated with antibiotics, but using a condom is an effective way to avoid contracting this infection.

Gonorrhoea: Gonorrhoea is another bacterial infection that is spread through sexual contact (vaginal, oral or anal sex). The infective bacterium is *Neisseria gonorrhoeae* (after the German bacteriologist, Neisser). Half of infected women and one in ten infected men have no symptoms. This is disturbing, as the later consequences, such as pelvic inflammatory disease in women, can be very serious. Sometimes gonorrhoea does cause symptoms, including a yellow or green discharge from the vagina or from the male urethra, painful urination and irritation or discharge around the anus. It can also be passed from an infected mother to her baby, as the baby is born, when it may infect the baby's eyes. It is treatable with antibiotics, although it seems that some resistant strains are emerging. Prevention is better than cure and using a condom is over 99 per cent effective in stopping the transmission of infection.

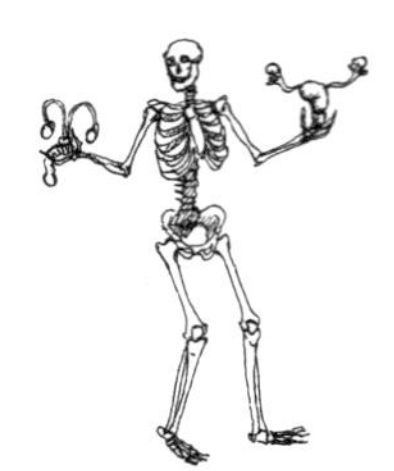

Herpes: Herpes infection is the most common cause of ulcers in the genital areas. *Herpes simplex* is a virus that causes both cold sores and genital infections. Genital herpes sometimes causes only subtle symptoms, such as being a bit itchy or red around the genitals. Sometimes it causes general, flu-like symptoms – a feeling of tiredness, aches and swollen lymph nodes. It can also cause blisters on the genitals, which burst to form ulcers, although these usually heal within two weeks. But this doesn't mean the virus has gone: it lies dormant in nerves and may recur when you're stressed. Herpes can be treated with antiviral drugs, but it's best to avoid it in the first place. Barrier methods like condoms work well, but if you or your partner have symptoms, it's important to avoid sex as the virus is very contagious – and go and see a doctor!

HPV: HPV – the human papilloma virus – causes genital papillomas or warts. Genital warts are the most common STD diagnosed in the UK. Although only one per cent of people with HPV develop genital warts, there are other, more worrying consequences of infection with the virus: almost all cases of cervical cancer are associated with HPV infection (see page 167). Vaccines are being developed to tackle HPV and reduce the risk of cervical cancer.

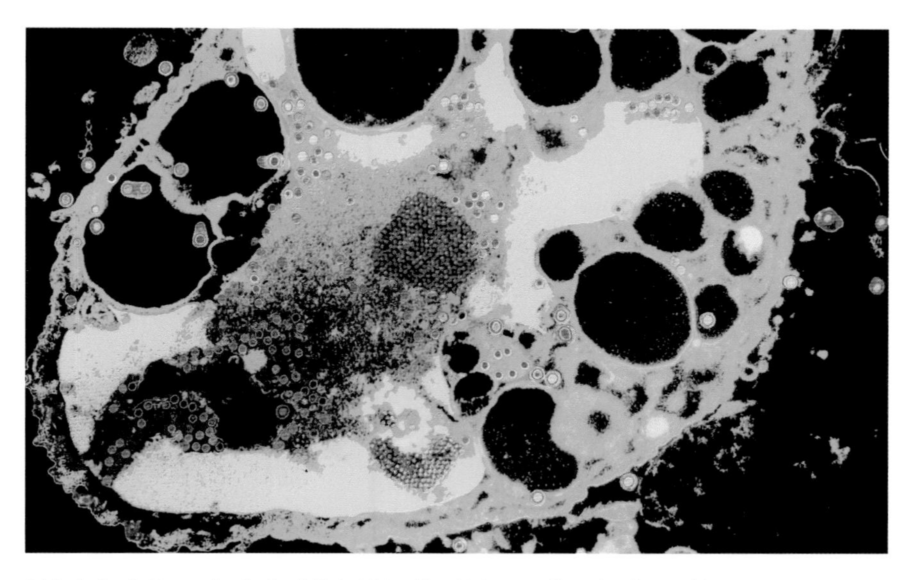

A hijacked cell: *Herpes simplex* has infected this cell and taken over its molecular machinery, commanding it to make more viruses (the red and pink spots), ready to infect the next cell.

Syphilis: At one time, syphilis seemed to have disappeared into the realm of 'horrific historical diseases'. But it's back, and with a vengeance. Syphilis is a really nasty disease: the initial infection causes rashes, swollen lymph glands and genital ulcers. The bacteria can then lie dormant, and later reappear to cause ulcerating lumps, erode into bone, cause cardiovascular diseases and a horrific brain condition involving dementia and seizures, which still goes by the chilling name of 'General paralysis of the insane'. Syphilis can also be passed on from mother to baby. It's caused by a strange little spiral bacterium called *Treponema pallidum*.

If it's a disease that is preventable and treatable with antibiotics, why on earth is it on the increase? There was a reduction in syphilis in the 1980s, probably to do with a change in sexual behaviour, as more people adopted safe sex in response to AIDS. But since the late 1990s, syphilis has made a startling comeback: the incidence in 2002 was over seven times that in 1997. The major factor underlying this increase seems to be a rise in unprotected sex – including unprotected oral sex. Whether this happened through a lack of knowledge or through complacency, the message is clear: practising safe sex is of utmost importance if you want to avoid this disease.

Other infections: As well as being transmitted by a sexual partner, infections can also originate in your own body. In the same way that bacteria from the anal region can cause cystitis in women, they can also cause vaginal infections. Wiping front to back and keeping the genital area clean are basic, good ideas, but it's also best to avoid using perfumed bubble bath, soap and talcum powder on these areas. Cleaning your genitals (on the *outside* – vaginal douching should be avoided) with plain water is the best way.

Just as there are 'friendly bacteria' in the gut, there are also symbiotic microflora in the vagina. Lactobacilli in the vagina do a good job of fighting off potentially pathogenic bacteria like *Gardnerella* and *E. coli*. Without these friendly bugs around, the not-so-good-natured variety can get a hold. Bacterial infections can increase the likelihood of contracting other infections, as well as being linked to premature births. Thrush is a common non-bacterial vaginal infection, which often takes hold in women who

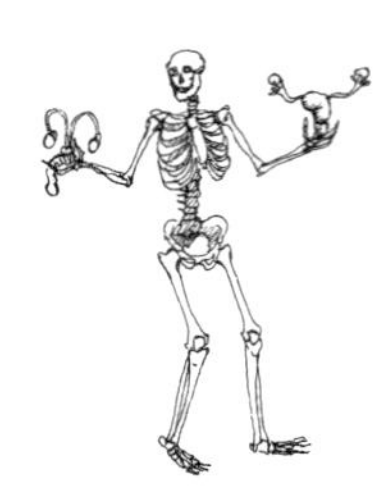

In Hogarth's *A Harlot's Progress*, Moll, a prostitute, is dying of syphilis while the doctors argue in vain about ineffective treatments.

have had a course of antibiotics. Thrush is a fungal infection caused by the fungus *Candida albicans*, giving the infection its other name: candidiasis. This fungus is often present in the vagina but doesn't cause any symptoms; a healthy vaginal flora seems to keep it under control. But it may turn into a symptomatic infection, with itching and discharge from the vulva and vagina, sometimes just before a period, especially if you take the contraceptive pill or if you've had a course of antibiotics (which will have killed your protective vaginal flora). Probiotics have been suggested as a possible way to prevent thrush; some studies have suggested that eating yoghurt might reduce the occurrence of thrush infections but others have shown no effect on the rate of thrush infections, even with vaginal *Lactobacillus* treatment.

In men, poor hygiene in the nether regions may cause a condition called phimosis in which the foreskin gradually tightens so that it can't be retracted, which can make urinating and having sex quite painful. Soaps and shower gels can irritate the foreskin and cause phimosis, so as for women, it is best just to use plain water.

This hasn't been a comprehensive tour of the infections you can get in your genital area and it's not meant to be a guide to self-diagnosis. If you've got pain, itching, soreness or lumps in the genital region, or a vaginal or urethral discharge – go and see your doctor sooner rather than later. You don't need to be embarrassed, because they're certainly not going to be; doctors see this sort of thing all the time.

Period pains

For many women, the normal monthly purging of the womb can be traumatic. For some (lucky) people, the loss of the lining of the womb each month is an insignificant event, easily mopped up with a small cotton tampon and completely pain-free. For others, day 21 of the uterine cycle is a day to be feared: premenstrual tension, debilitating cramps and apparently unbelievably heavy flows of menstrual blood are a regular feature from the teens to the menopause.

The posh word for these pains, which originate in the uterus, is dysmenorrhoea (Greek for 'bad monthly flow'). It's the most common gynaecological problem affecting women of all ages, all over the world. It has a huge effect on women's lives: half of all women have been absent at least once from school or work due to period pains. But very few women go and seek treatment for it.

It's also a condition that it has taken the medical profession an unfeasibly long time to take seriously! For years and years, dysmenorrhoea was considered to be a psychosomatic condition, complete with Freudian overtones relating it to anxieties about sex and learned behaviour perpetuating it through the generations from mothers to daughters. The Greek word for the uterus is *hysteros* (from which we get both the words 'hysterical' and 'hysterectomy'). But period pain absolutely isn't hysterical. There's a real, physical basis for dysmenorrhoea: as the lining of the uterus (the endometrium) is shed, endometrial cells release hormones – prostaglandins – which tell the uterine muscle to contract. Women with dysmenorrhoea have higher levels of prostaglandins, and therefore experience more painful cramps.

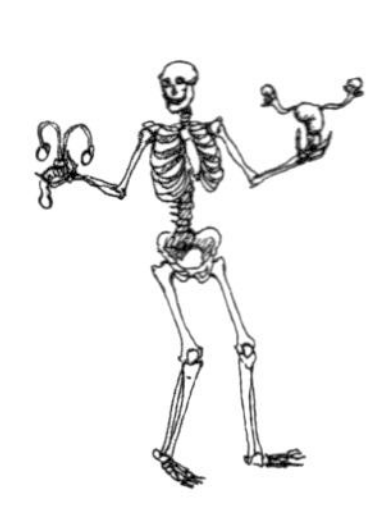

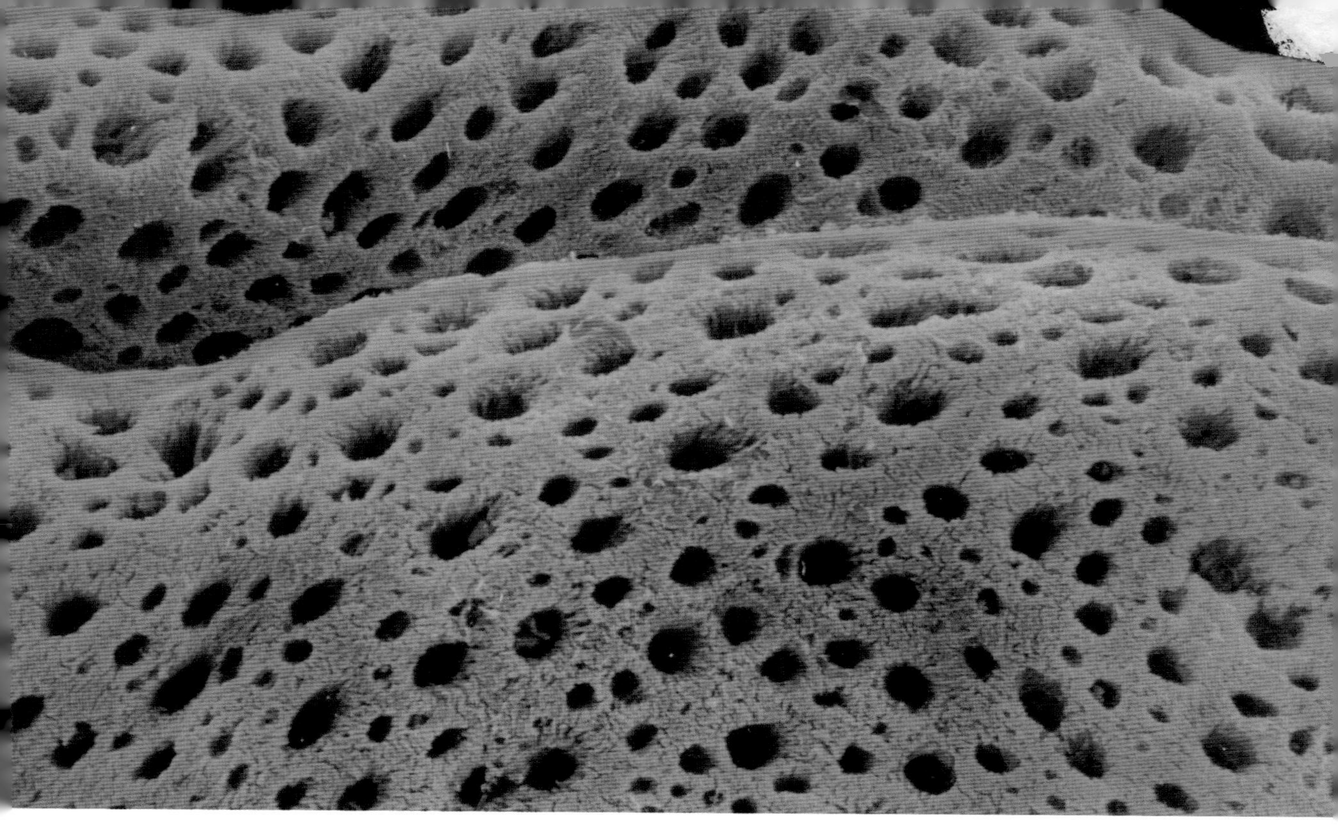

A comfortable cushion: the endometrium is at its thickest around the time of ovulation every month. It upholsters itself in anticipation of an embryo implanting, but if no embryo arrives, the lining is lost.

Period pain usually accompanies the onset of a period and typically lasts from eight to 72 hours. It can feel like cramp in the pelvis or the lower part of the belly, but it sometimes radiates out to other areas such as the back and the thigh. Some women get headaches, nausea and even vomiting. Period pains are very common and more likely if a woman started her periods at a relatively young age and if each period lasts a relatively long time. Those are factors you can't control but there are some you can: smoking, obesity and alcohol consumption all increase your likelihood of having dysmenorrhoea.

What about treatment? Painkillers are an obvious place to start. Aspirin and ibuprofen are part of the family of painkillers called non-steroidal anti-inflammatory drugs (NSAIDs). Their anti-inflammatory effect works to suppress prostaglandins, the hormones that cause the uterine cramps. NSAIDs are the first line treatment for period pain, relieving symptoms in 70 per cent of women but they can cause side-effects like nausea and vomiting, and must not be taken on an empty stomach (see page 87).

The contraceptive pill is an accepted treatment for period pains, although its usefulness is disputed; many studies have produced evidence to show that it works, but others have shown it to work no better than placebo. Having said that, if you don't want to become pregnant, the oral contraceptive pill is a useful option. The pill has received a lot of bad press, but most of the side effects are mild (headache, nausea, bloating). Serious side effects – like DVT, heart attack and stroke – are rare, but more common in women who smoke whilst on the pill. The pill also has its good side: as well as potentially improving period pain and preventing pregnancy, taking the pill lowers your risk of uterine and ovarian cancers.

Analgesics and the pill are common treatments, but your doctor should be able to give you good advice on the range of treatments available. If you've still got pain, or if you can't take these treatments for some reason, you might want to try something else. But which alternative remedies work?

Evening primrose oil is one alternative remedy that has a large following of women who swear by its ability to banish premenstrual tension. No large studies support this claim but this doesn't necessarily mean that it doesn't work and (even if it's the placebo effect doing its wonderful work), if it does work for you, then you might as well use it. Very few adverse effects of evening primrose oil have ever been reported. However, you should be very wary about taking evening primrose oil if you're epileptic, as it can cause seizures in some people.

It seems that taking vitamin B1 regularly may reduce painful periods: in one large trial, nearly nine out of ten women taking 100 milligrams of vitamin B1 daily were cured of period pain after two months. There's also some evidence that magnesium, pyridoxine (vitamin B6) and fish oil may be effective but more research needs to be done on these possible treatments before any definitive conclusions can be reached.

Transcutaneous electrical nerve stimulation (passing a tiny electric current through the skin to stimulate the nerves beneath), acupuncture and heat (the good old hot water bottle is vindicated!) have all been shown to help relieve the pain of menstrual cramps, but spinal manipulation appears to be no better than a placebo.

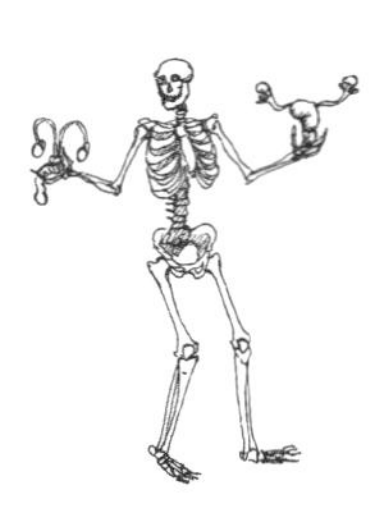

Diet and lifestyle also play a part. Obesity is certainly a factor, but on the positive side, exercise and a healthy diet may help combat period pains. One small study showed that a low-fat vegetarian diet helped to reduce symptoms of period pain, though the number of people involved was too few for it to be conclusive. Exercise might help by improving blood flow in the pelvis and also by releasing endorphins, the body's natural painkillers.

Erectile dysfunction

More and more evidence is stacking up showing a link between erectile dysfunction and heart disease, high blood pressure and diabetes. At the root of all these problems is a problem with the lining of blood vessels. Sometimes, erectile dysfunction might be the first clue that there's something rotten in the state of your blood vessels: a harbinger of heart disease. The lifestyle changes that lower your risk of heart disease should, logically, also lower your risk of getting erectile dysfunction: like exercise, losing weight and stopping smoking. Is there any hard evidence for this? Can lifestyle changes really help to overcome erectile dysfunction once it has occurred?

Smoking appears to double the risk of erectile dysfunction and unfortunately, stopping smoking in mid-life doesn't seem to remove this risk (so the earlier you stop, the better, really). Some studies have linked excessive alcohol consumption to erectile dysfunction, but the link doesn't hold up in the long term. At the moment, the effect of alcohol on erectile function remains unclear.

Obesity, a diet high in fat and a sedentary lifestyle are risk factors for developing erectile dysfunction. The most physically active men have the lowest risk of erectile dysfunction. Studies have shown that men who change their lifestyles to lower their risk or to reduce the impact of erectile dysfunction, see improvements in their cardiovascular health, their mood and their quality of life generally. To misquote a phrase that is usually associated with the mind (and I'm absolutely not accusing men of thinking with their reproductive organs!): *penis sana in corpore sano* – a healthy penis in a healthy body.

Gynaecological cancers

Breast cancer: In women who don't smoke, the most common cancer is breast cancer. This doesn't mean that you're more likely to get breast cancer if you don't smoke; it just means that you've sidestepped lung cancer and breast cancer is next in line. It is the most common cause of death for women between 40 and 50, accounting for about one in five deaths. Screening programmes in the over-fifties and improved treatment have increased the chances of survival. But is there anything you can do to lessen your risk?

Some factors you can't do much about: your age, having someone in your close family who's had it, how old you were when you started your periods, and the age you were when they stopped. Taking the pill very slightly increases your risk, but you obviously have to balance that against the risk of an unwanted pregnancy. Hormone replacement therapy also slightly increases the risk of breast cancer, if taken for more than five years (but in short-term treatment for menopausal symptoms, the benefits are likely to outweigh the risks; discuss this with your doctor).

There are other factors that you can do something about. In post-menopausal women, obesity is linked to an increased risk of breast cancer. Oestrogen is made in fat tissue; the fatter you are, the more oestrogen your body will be making. The higher level of oestrogen contributes to a higher risk of breast cancer. Drinking too much alcohol is also a contributing factor, whereas eating plenty of vegetables reduces the risk.

Ovarian cancer: Ovarian cancer, though less common than breast cancer, causes about three per cent of deaths from cancer, across all ages (compared with twelve per cent for breast cancer and two per cent for cervical cancer). There is convincing evidence that long-term use of the oral contraceptive pill, and having babies, lowers your risk of developing ovarian cancer. Some studies have suggested that there is a link between talcum powder and ovarian cancer, but the evidence is circumstantial and inconsistent. Talcum powder has certainly been found in women's ovaries (the oviducts are open and talcum powder can get up the vagina, through the uterus and oviduct and reach the ovary). Some of the ovaries that have been found to contain

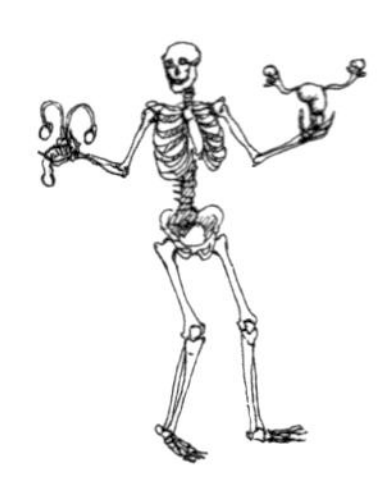

talc were cancerous whereas others were normal. Of the millions of women who use talcum powder, only a very few get ovarian cancer.

Cervical cancer: It's not often you get to see an eponym. At the Bristol University degree ceremony in 2006, not only was I lucky enough to watch my students graduate, I also saw Professor Epstein being awarded his honorary doctorate. Tony Epstein gave his name to the Epstein-Barr virus.

Epstein-Barr virus causes glandular fever. But in the 1960s, Tony Epstein spotted the link between this virus and a cancer of the jaw. For the medical community of the time, this was hard to stomach; it just seemed too weird that an infection could cause cancer. However, the evidence was compelling and further investigations have produced even more evidence for the link. Since then, many links have been found between viral infections and cancers of many different tissues of the body. One of these is the link between infection with the Human Papilloma Virus (HPV) and cervical cancer. Cervical cancer is almost always associated with HPV infection. It would seem sensible to try to avoid catching HPV, but this may seem easier said than done as there's usually no sign that a man might carry the virus. The best thing to do is play safe. Using a condom prevents transmission of HPV and many of the other venereal infections that are looking for a new victim.

What you can do: Your diet and lifestyle have a large impact on your risk of developing a gynaecological cancer. Contracting HPV increases the risk of cervical cancer; smoking increases the risk of developing breast, ovarian and uterine cancers (and others). Obesity is a risk factor for gynaecological cancers and other types of cancer. (In the western world, smoking causes nearly a third of deaths from cancer and obesity another third.)

These are statistics at the society level. To put it into a more personal perspective, an obese woman has a 60 per cent higher risk of dying from cancer than a woman of normal weight. Eating fruit, vegetables and antioxidants reduces your risk of developing cancer; eating a diet with lots of animal fat increases your risk. Exercise is a positive influence: regular physical activity protects against ovarian, uterine and breast cancer.

Prostate and testicular cancer

Prostate cancer: The prostate is a chestnut-sized gland that sits just beneath the bladder. After the urethra leaves the bladder, it passes through the prostate gland, where it is joined by the vas deferens on each side. The urethra plays a dual role in men: it forms a passageway for conveying both urine and semen out of the body, whereas women's urinary and reproductive systems are, of course, more highly evolved and kept separate. (This is tongue in cheek! Men are just as 'highly evolved' as women. It's just that primitive mammals had a single hole for everything – faeces, urine and sex – which gradually got divided up into separate conduits.)

The prostate gland is part of the reproductive system – it empties its secretions into the urethra to join the sperm just before ejaculation. The prostate tends to grow larger with age, resulting in a condition called benign prostatic hypertrophy (BPH), which is a non-cancerous over-growth of the prostate. An enlarged prostate can block the urethra, making it difficult to pee. This is a common problem and severe cases can be overcome with surgery to widen the part of the urethra running through the prostate.

However, malignant cancerous growths can also start in the prostate. Prostate cancer is the most common cancer in men. It can cause problems peeing, just as BPH does, which is why it's a good idea to consult your doctor if you're experiencing problems like this. There isn't an easy way out; there's no pill you can take to eliminate your risk of prostate cancer. Changes to your lifestyle far outweigh the effects of any dietary supplements when it comes to reducing your risk of death from prostate cancer. What's good for your heart is good for your prostate; keep your heart healthy (see page 69) and your prostate should be happy too.

Testicular cancer: Testicular cancer, compared with prostate cancer, is quite rare. But it's still a good idea for men to check themselves – and the testes, unlike the ovaries, can be examined easily. You need to get used to the normal feel of your testicles. Remember, there's supposed to be a lumpy bit at the back (the epididymis), extending up into what feels like a cord: the vas deferens. If you feel any new lumps and bumps, do go and see your doctor.

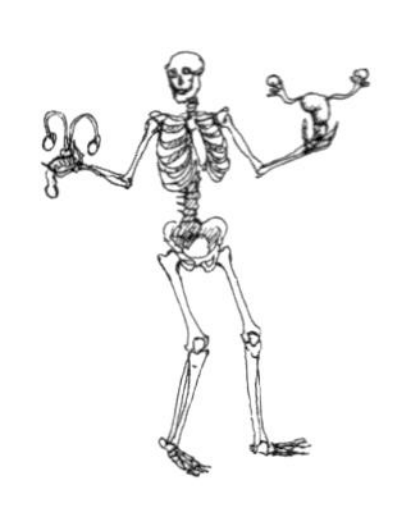

Five ways to keep your reproductive organs healthy:

- Practise safe sex. Using a condom significantly reduces your risk of catching a sexually transmitted disease.

- Keep your genitals clean on the outside: wash with plain water, not perfumed soap or bubble bath; women should make sure to wipe from front to back; empty your bladder as soon as possible after sex.

- Never leave a sanitary tampon in for longer than six hours; leaving a tampon in for longer increases your risk of vaginal infections.

- Don't smoke – it increases your risk of all sorts of cancer.

- Keep active, keep your weight down and eat healthily to reduce your risk of developing cancer.

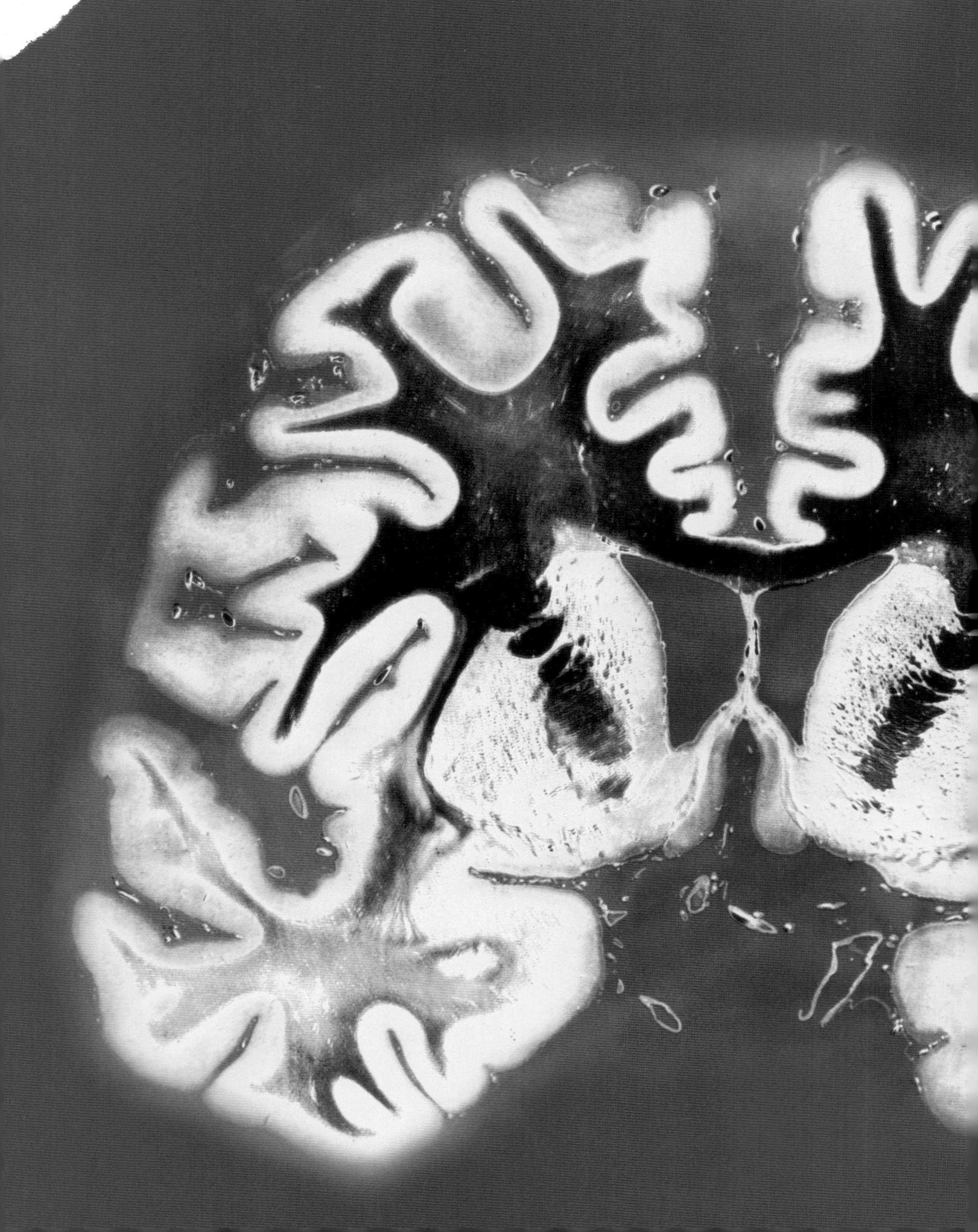

EIGHT

THE BRAIN

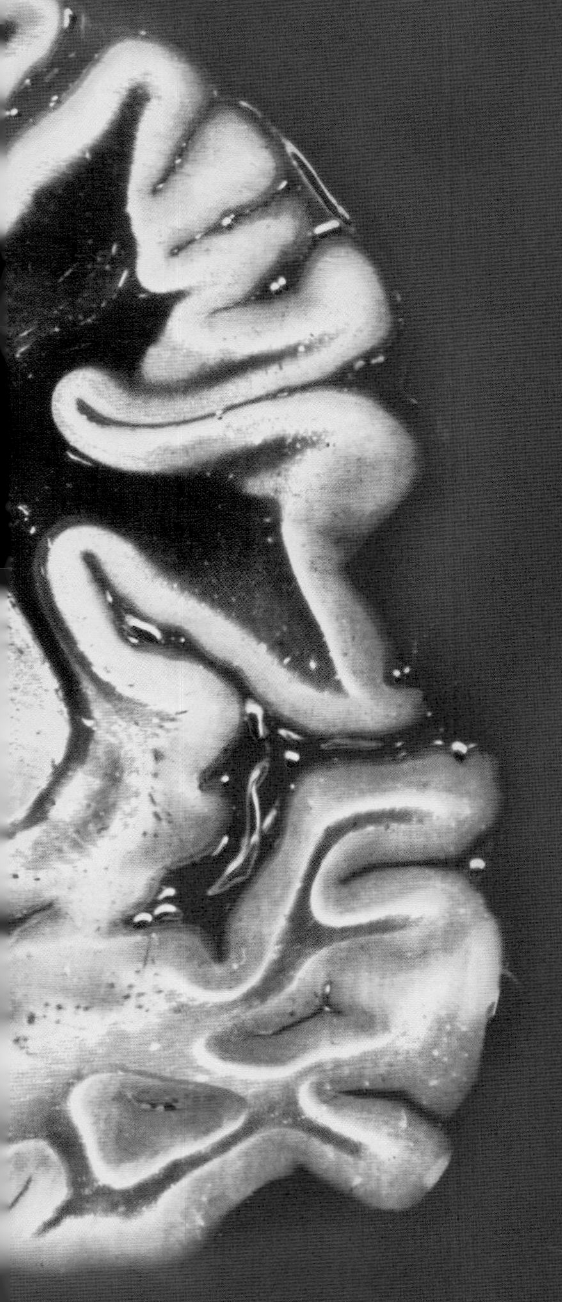

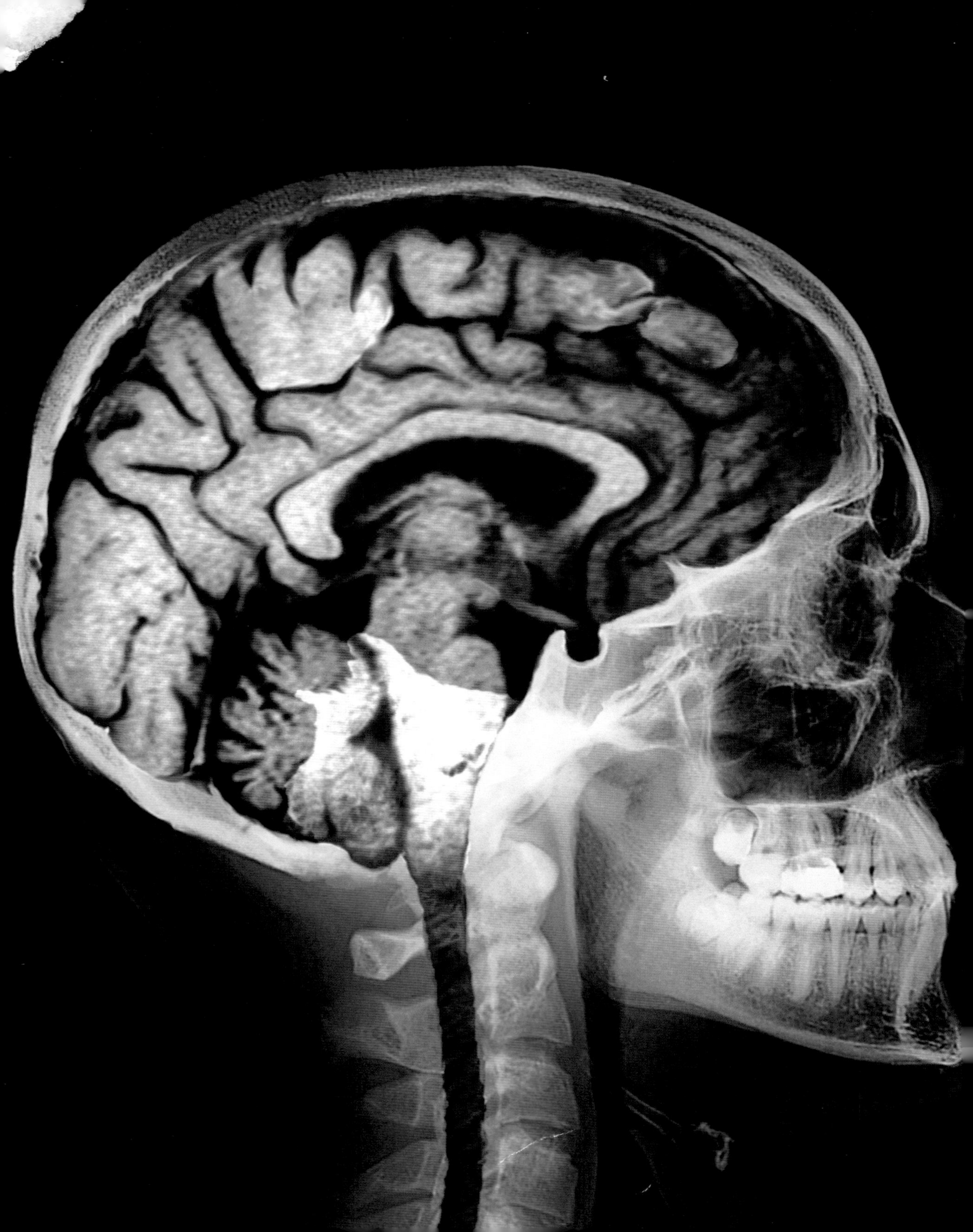

The human brain is the most complex thing we know. It looks quite simple: a lump of pinkish-greyish stuff, but inside it contains about 100 billion neurons (nerve cells) with a million billion connections between them. We're only just beginning to understand the brain properly: advances in computing are allowing us to create models of how the brain might work, but we're still a very long away from knowing exactly what's going on up there.

The brain is well protected inside the hard skull but this very protection can become a problem if something, such as a tumour or bleeding, increases the pressure inside the skull. All parts of the brain need a good, constant blood supply: arteries narrowed with atheroma are bad news for the brain, just as they are for the heart. Stroke – the equivalent of a heart attack in the brain – is the third most common cause of death in developed countries.

Previous pages **In this false-colour section through a brain, white matter is brown, whilst the grey matter of the cortex is shown in yellow.**

Left **Mission control: snugly fitted inside the skull, the brain sends messages out and gets information back from the rest of the body via the spinal cord.**

Where is the brain?

Open up your skull, and you will find the brain sitting there, wrapped in several layers of membrane known as meninges. A thick, almost leathery membrane lines the inside of the skull: this is the dura mater. (This strange term, meaning 'hard mother', entered anatomical language as a translation of an Arabic term into Latin. In the original Arabic, the 'mother' bit indicated a physical closeness or relatedness between two things.) The dura is lined with another membrane, the delicate arachnoid (Greek for 'cobweb') mater. The brain itself is covered in a very thin membrane, the pia (Latin for 'tender') mater. Between the arachnoid and the pia is the subarachnoid space, which is filled with cerebrospinal fluid (CSF). This fluid originates deep inside the brain in spaces, or ventricles, containing lumpy networks of blood vessels, which produce CSF. This fluid then flows out of a canal at the back of the brain and into the subarachnoid space, bathing the outside of the brain. The subarachnoid space stretches all the way down the spine, so CSF also bathes the spinal cord.

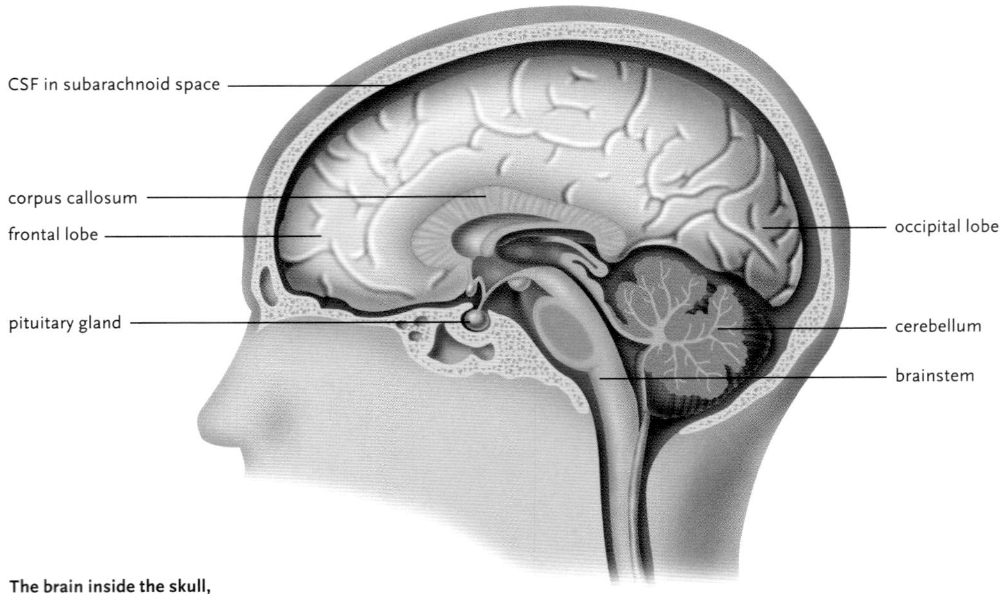

The brain inside the skull, surrounded and filled with CSF.

CSF provides some cushioning for the brain, and transports nutrients and waste products. The composition of CSF can provide clues in diseases of the brain and spinal cord. It can be sampled by pushing a needle into the subarachnoid space in the lower spine, called a 'lumbar puncture'.

What does your brain look like?

I have to say, from an anatomist's point of view, the brain is a pretty ugly and uncharismatic organ, but then I'm not a neuro-anatomist (yes, there are people who spend their lives concentrating on investigating the structure of the brain). The cerebral hemispheres look for all the world like a large walnut, almost split into a left and right side. Tucked underneath the cerebral hemispheres is a rippled bulge, the cerebellum. The inside of the cerebellum has an interesting tree- or fern-like pattern. Right at the bottom of the brain is the brainstem, joining the brain to the spinal cord.

Cut through the brain and you find a grey rind (the cortex), with a white core beneath it and then more grey lumps in the middle. I find it utterly remarkable how unremarkable-looking the brain is. But there it is – the most complex organ in the body and the home of your mind. It's difficult to come to terms with the idea that this physical thing holds the very essence of you.

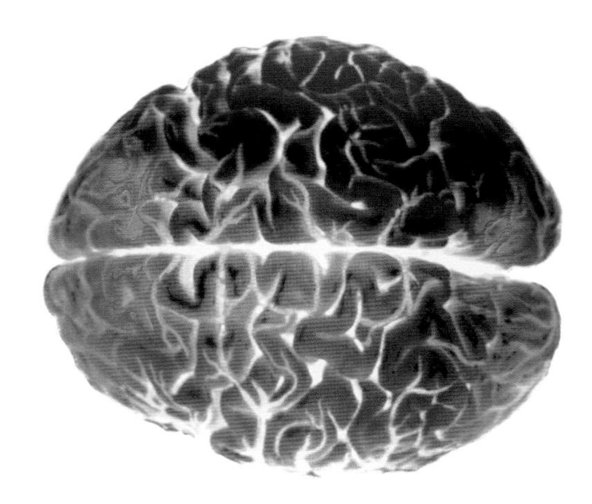

Like a walnut: on the surface of the brain, the ridges (gyri) of the cortex are separated by grooves (sulci).

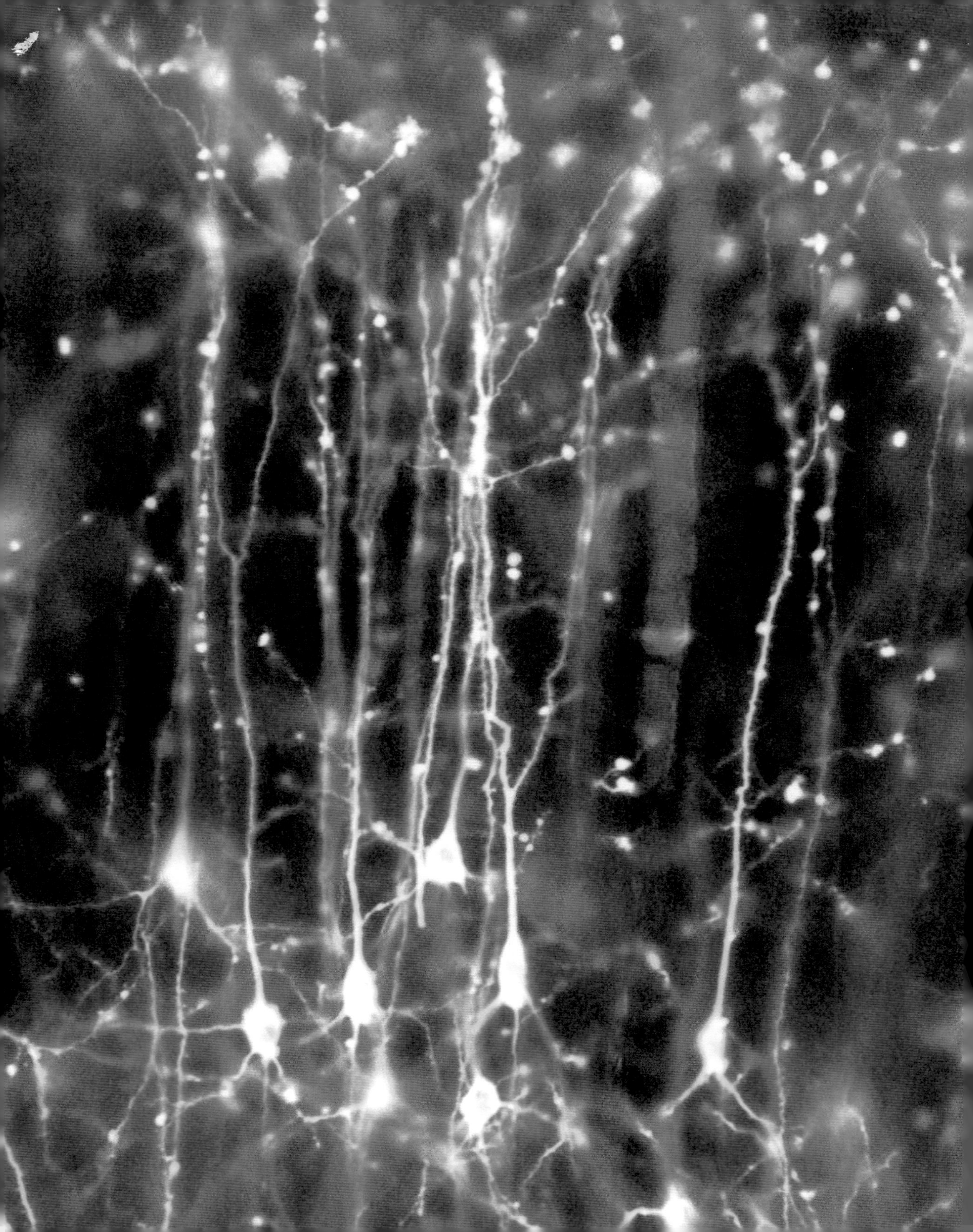

How the brain works

When you get down to a much smaller scale, the brain starts to reveal some of its secrets. Nerve cells (neurons) are very odd-looking cells. They have a small cell body, containing the nucleus and all the other machinery of the cell, and this cell body has lots of long, narrow, wiry things poking out from it. Those wiry things are exactly that: electric wires. One of these is normally longer than all the rest, and this is called the axon (Greek for 'axis'), and carries nerve impulses away from the cell body. Within the brain, axons may be very short, even less than a millimetre long. But nerve cells in the spinal cord which send messages to muscles can have extremely long axons: nerves which go from the lumbar region (at the base of your spine) to your toes are bundles of axons which are each up to a metre long. Nerve cell bodies also have shorter wires called 'dendrites' (Greek for 'of a tree'). Inside the brain, all these wires form connections with wires from other nerve cells, forming a massive network of neurons.

Nerve impulses – electric currents – travel along the axons of neurons as waves of electrically charged ions move across the cell membrane. Some axons are wrapped up in a coat of fatty cells, the myelin sheath. This sheath stops the current leaking out of the axon, just like the plastic coat insulating the outside of a copper wire.

Nerve cells aren't joined together. There is a very tiny gap – a synapse – between the axon of one cell and the dendrite of another. When the electrical current reaches the end of the one neuron, it somehow has to jump the gap to the next cell. The current makes the end of the axon release chemicals – neurotransmitters – into the synapse. The neurotransmitter switches on the next neuron, generating an electrical current. There are many different types of neurotransmitters, including adrenaline, noradrenaline, dopamine and glutamate. Glutamate is the main 'switching on' neurotransmitter. However, too much glutamate may damage neurons, and an excess of this is thought to lie at the very root of several brain and nerve diseases.

Cranial constellation: pyramidal neurons (the cell bodies, at the bottom, are shaped like pyramids or pears) in the cortex of the brain, with their long stalks (axons) projecting upwards.

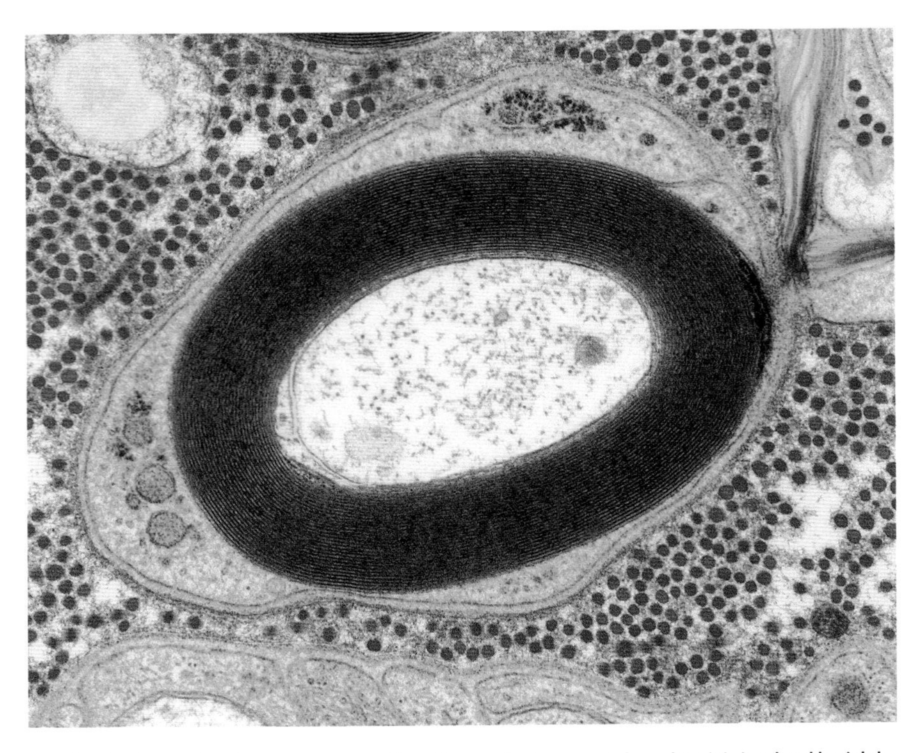

Electrical insulation: this slice across an axon (the white oval in the centre) shows how it is insulated by tightly-wrapped layers of myelin (brown) made by the Schwann cells (green); magnified over 14,000 times.

Regions of the brain

Although the brain may look symmetrical on the outside, it's far from it on the inside. For one thing, the right side of the brain controls the movement of the left side of the body and the left side of the brain controls the right side of the body. Brain functions are divided between the two hemispheres: in most people, the right hemisphere is the creative side; it 'does' spatial perception, art and music. The left hemisphere is usually the dominant side; where control of language and logical thought reside.

What goes on precisely where in the brain is a question that has taxed neuroscientists for over 100 years. In the Victorian era, the idea that certain areas of the brain did different things was elaborated into the pseudo-science of phrenology: practitioners of this craft claimed to be able to analyze personality by examining the shape of the head. This is, of course,

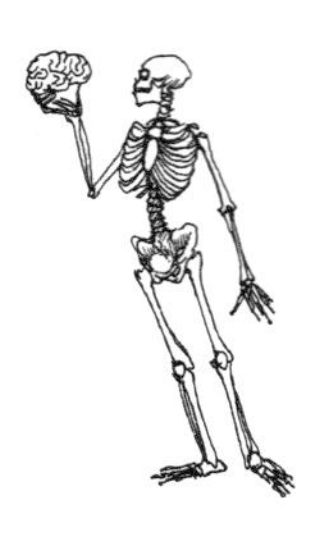

completely absurd. However, patterns of impairments that arise when areas of the brain are damaged show us that different functions are somehow localized in the brain. For example, damage to the frontal lobes can cause intellectual impairment or personality changes, damage to the left temporal lobe can wreck a person's numeracy or writing skills, and damage to the right temporal lobe can stop someone being able to recognize faces.

It's clear that there are regions to be mapped in the brain – it's just a question of working out how to draw the map. With the aid of new imaging technology like functional magnetic resonance imaging (fMRI), neuroscientists are beginning to be able to work out precisely which bits of the brain are involved with different thoughts, emotions and actions. The trouble is, it seems that the network-like nature of the brain means that, rather than neurons used for one particular task being packed closely together in a discrete area, they are spread over the brain. There's a lot more work to do before we know our own minds.

Aspects of personality are mapped onto the head in this useful Victorian diagram of a phrenology head – but don't bother checking your head: personality isn't evident from the head, the palms, or any other bits of us on the outside – it resides firmly inside the brain.

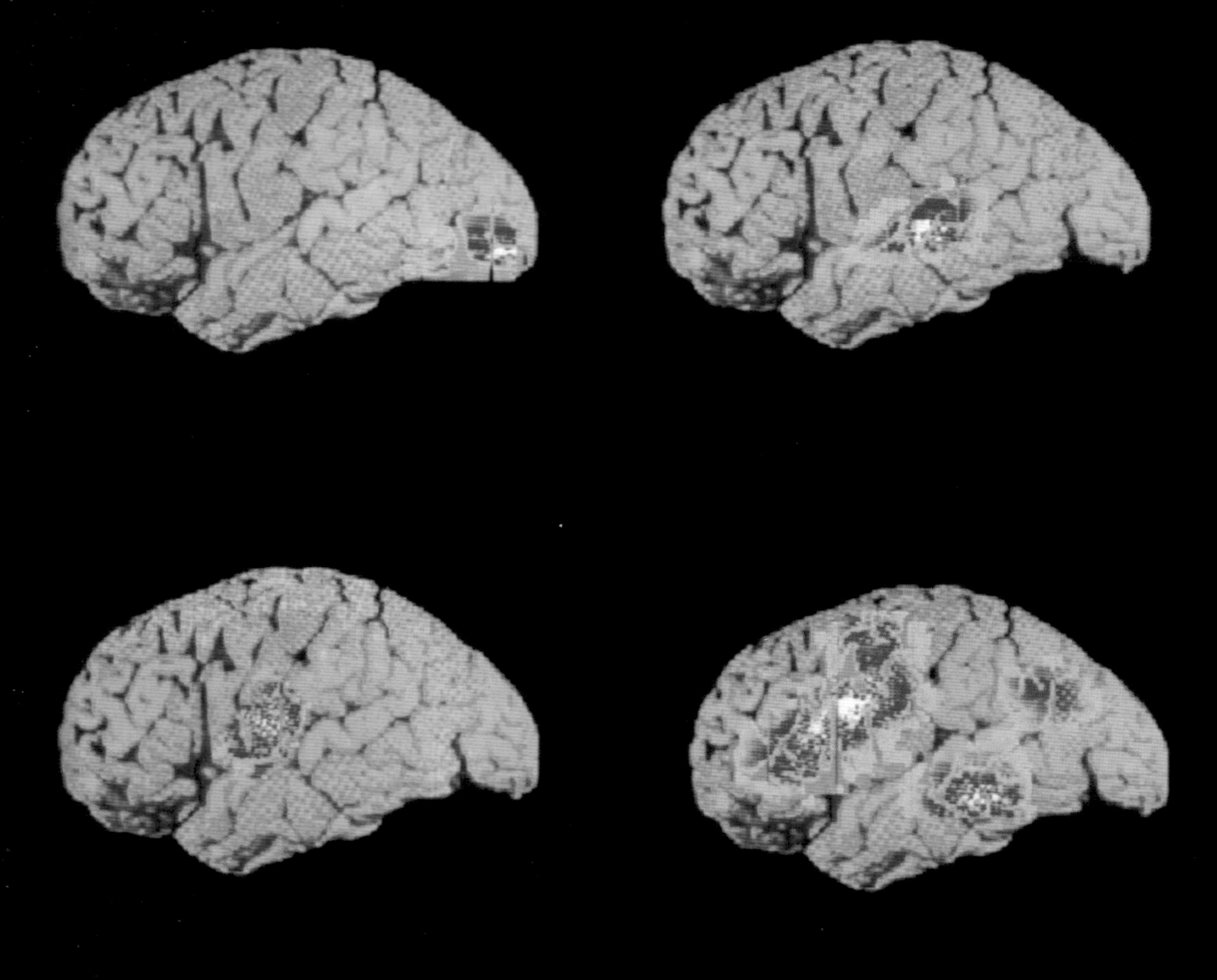

The brain working: using fMRI, some areas of the brain light up with particular tasks; clockwise from upper left: sight; hearing; *thinking* about words; speaking.

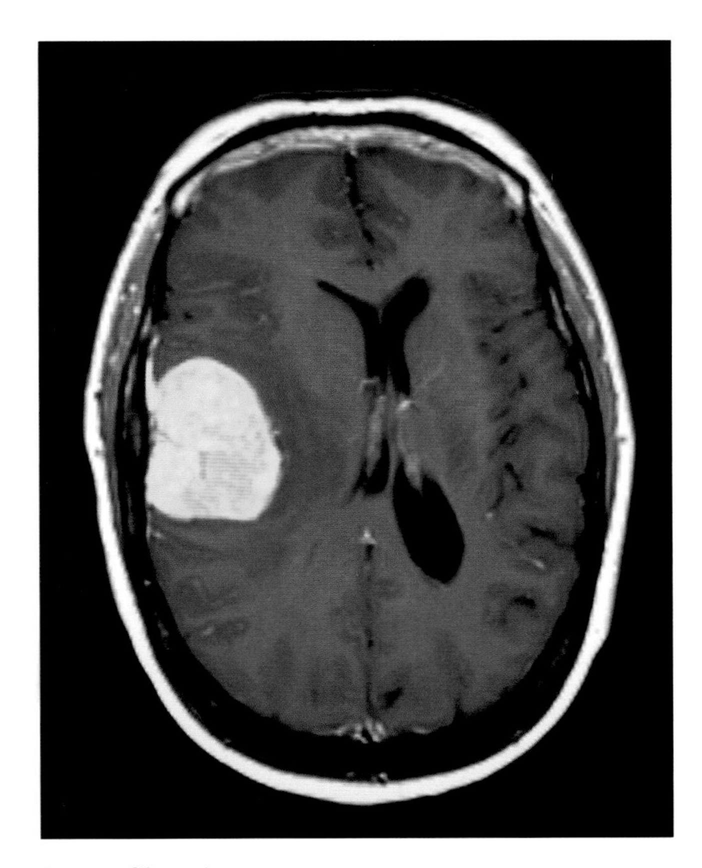

A tumour of the meninges (yellow) is taking up precious room inside the skull, squashing the brain.

looking for the tell-tale sign of a blurred optic disc on the retina at the back of the eye.

By and large, damage to the brain results in one of two outcomes: neurons might get irritated and 'jumpy', causing waves of electrical impulses in the brain, producing epilepsy, or the neurons may be destroyed, and their particular function lost.

In terms of looking after the brain, it really is a question of *mens sana in corpore sano*: a healthy mind in a healthy body. A balanced diet with plenty of fresh fruit and vegetables will keep the brain well-nourished. Like other bits of the body, the brain likes regular exercise – mental exercise. Smoking, you won't be surprised to hear, is bad for the brain. And though we don't fully understand what the brain is doing when we sleep, a good night's sleep is absolutely essential to keep the brain working at its best.

I'll now look at how you might reduce the risk, or the effect of, some common conditions affecting the brain.

Headaches

The exact cause of headaches remains shadowy and obscure. It's sometimes touted as being to do with irritation of nerves and vessels or tension in the muscles of the scalp. Headaches are very diverse: some are sharp and piercing, others dull and throbbing; some feel like a tight band around the head, and others like a pressure behind the eyes. Headaches can be brought on by a range of external and internal influences: depression, minor head or

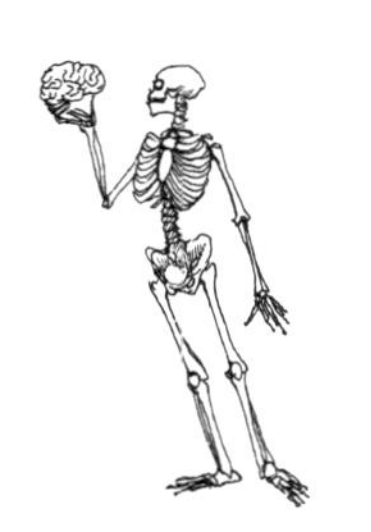

Common brain problems, and how to prevent them

Damage to the brain can come from the inside or the outside. High speed acceleration or deceleration, as happens in traffic accidents, can cause tearing of the nerves. Direct head trauma (such as blows to your head or your head hitting something) can cause damage to neurons, swelling or bruising of brain tissue or haemorrhage (bleeding) inside the skull – even if the skull isn't broken. In a mild traumatic brain injury, the person feels initially dazed but quickly recovers. In more severe injuries, there may be a period of amnesia (memory loss) which can last weeks, or the person may go into a coma, a profound state of unconsciousness.

From the inside, the brain can be damaged by tumours, haemorrhages and clots. Brain tumours and haemorrhages have direct effects; for example, tumours can invade other parts of the brain, but also compete for precious room in the compact and *bijou* accommodation that the brain occupies inside the skull. The brain sits in a tightly packed compartment; tumours and haemorrhages cause rises in pressure, which can have effects throughout the brain, crushing neurons and cutting off blood supply.

Because the lining of the optic nerves (the nerves that take impulses from the eyes to the brain) is continuous with the meninges lining the brain, a rise in pressure in the cerebrospinal fluid around the brain will be transmitted along the optic nerve. Within the eye, the optic disc, which is where the optic nerve leaves the retina, bulges. This means a doctor can check very simply and quickly for a rise in intracranial pressure by

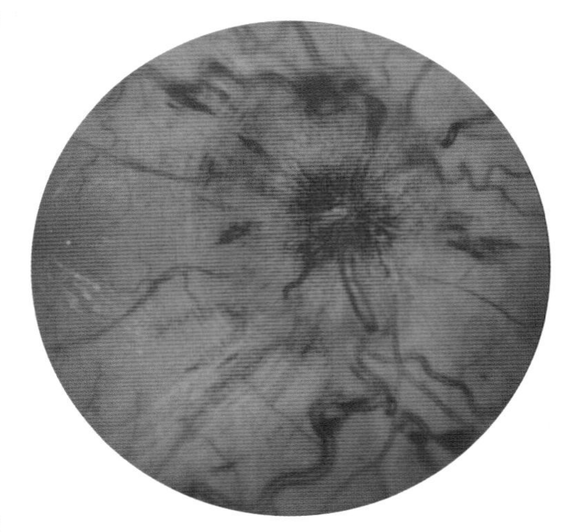

Papilloedema: the optic disc on the retina should look like a yellow doughnut, but here the disc is just a reddish blur (upper right), indicating high pressure inside the skull.

neck injuries, stress and anxiety or eye strain. The vast majority of headaches are essentially harmless. They may hurt but there's no underlying damage to worry about. Stretching, massage and simple painkillers such as paracetamol or aspirin may help relieve the pain and tension. Although most headaches do clear up by themselves, it's important to remember that headaches can occasionally be signs of something more serious. If the pain is unremitting, you have persistent blurred vision, you feel sick or you find yourself getting recurring headaches and having memory problems – go and see your doctor.

Migraines

The word 'migraine' is a strange one – it seems to come from the Latin word *hemicrania*, meaning 'half-head': literally, a splitting headache. The boundary between 'ordinary' tension headaches and migraine is difficult to define. The headache of migraine is associated with disturbances to both vision and stomach. Some migraine headaches are heralded by visual disturbances, such as dark patches in your vision, flashes of light or jagged lines like battlements, and sometimes by tingling and weakness on one side of the body. Early migraine symptoms can be similar to a transient ischaemic attack (see pages 185–6), with a feeling of weakness on one side of the body. You should treat such symptoms as an emergency unless you've had them before and have been reassured that they are part of a migraine. This early phase can last minutes to an hour, then the headache crashes in. During migraine, blood vessels in the brain and the meninges get wider, the tissue swells and sensory nerves are stimulated. Although migraine can cause severe pain, this doesn't mean that any real damage is happening inside the brain.

There's a recognized link between factors such as relaxing at the end of the week, chocolate, cheese, noise, lights, PMT, and migraine – but how these factors set it off is a bit of a mystery. Although avoiding foods that set migraine off might seem logical and sensible, it doesn't actually seem to help much. Painkillers like paracetamol may help, but repeated use may also cause more migraines. There are other drugs that can be used for severe migraine – ask your doctor.

Stroke and TIA

One in nine deaths in the UK is from a stroke. A stroke happens when the blood supply to an area of the brain is interrupted; just like the starved heart muscle in a heart attack, that area of the brain dies. In eight out of ten cases, the blood supply is halted by a clot lodging in a narrowed artery, just like a heart attack, although some strokes are caused by bleeding inside the brain.

The brain is supplied with oxygen-rich blood through two pairs of arteries: the internal carotid and the vertebral arteries. The internal carotid arteries run up the neck, through a channel in the base of the skull and pour their contents into a circular artery called, rather grandly, the Circle of Willis. The vertebral arteries travel up the neck, through small holes in the sides of the vertebrae. When they get to the skull, they disappear inside the big hole in its base, the foramen magnum. Inside the skull, the vertebral arteries join up to make a single artery which runs up under the brainstem. This also ends in the Circle of Willis, from which arteries then branch off to

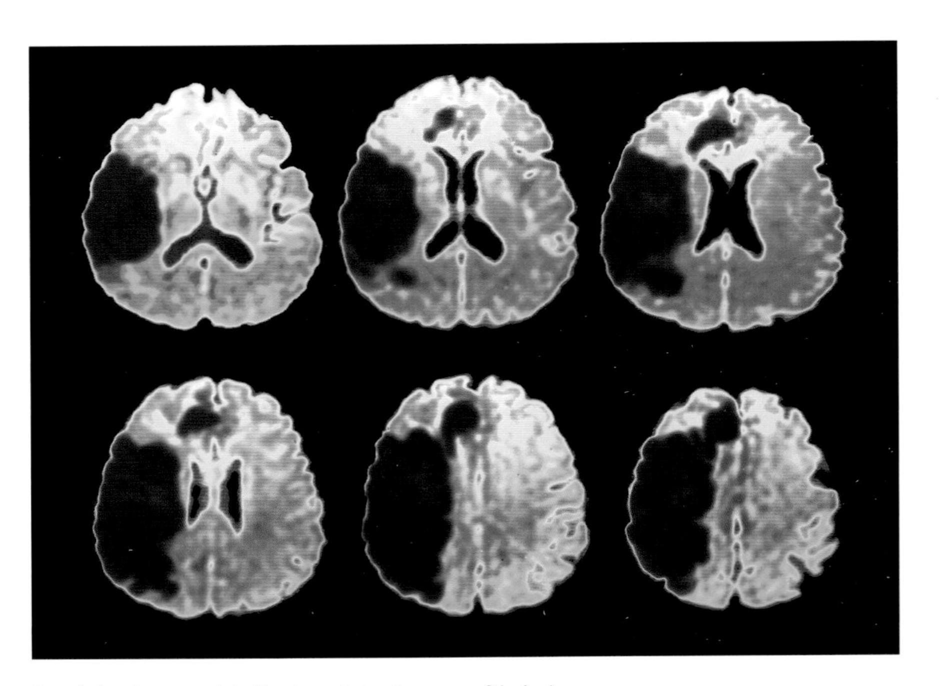

A stroke has interrupted the blood supply to a large area of the brain, which appears black on this coloured MRI scan.

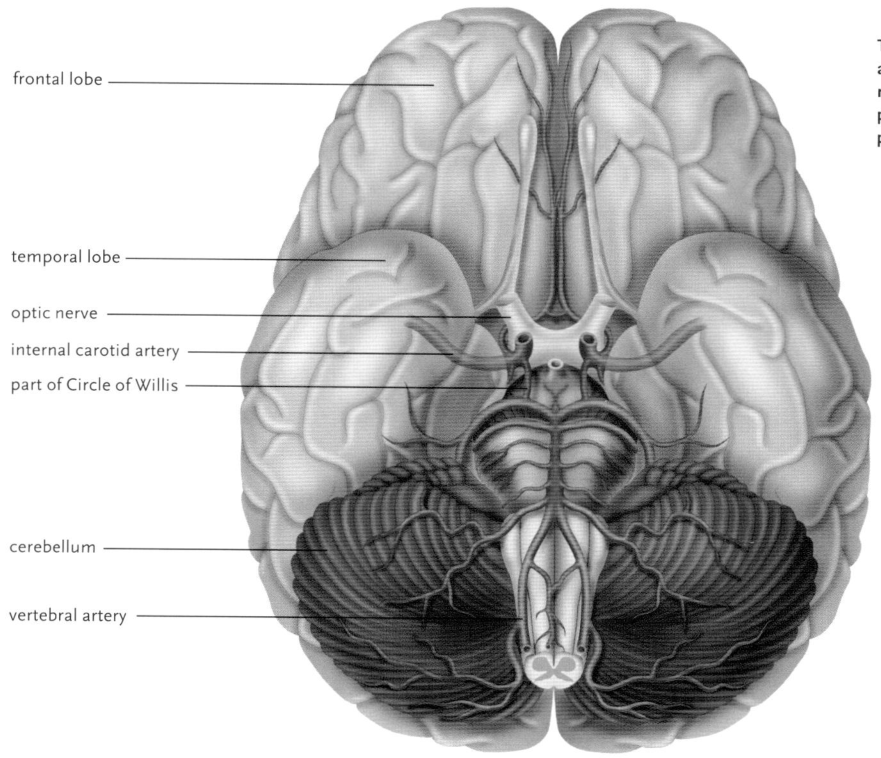

The Circle of Willis: forming a ring around the optic nerves, this circle of arteries provides branches to all parts of the brain.

supply different parts of the brain. This roundabout-like vessel is a clever design – a safety device that means that if any one of the incoming arteries is blocked, blood from the remaining arteries can still be distributed to all parts of the brain. However, the cerebral arteries leading from the Circle of Willis also get blocked. A clot obstructing a large vessel in the brain can cause the death of a significant area of brain tissue and a stroke ensues. This will be fatal in 25 per cent of cases. A survivor of a stroke may be left with a loss of sensation or weakness in one side of the body, problems with speech or other difficulties.

Sometimes the blockage of an artery in the brain is partial and temporary, producing a transient ischaemic attack (TIA). This may be noticed as a passing weakness in an arm or leg, momentary blindness in one eye or problems with speaking or understanding language (aphasia). A TIA may pass within seconds or last up to 24 hours. But TIAs are far from harmless. If you have one, you're likely to have another and they may be omens before a full-blown stroke. TIAs and strokes are medical emergencies – if you think you or someone else is having one, you should call an ambulance.

Some very small blockages in vessels in the brain may go unnoticed. But over time, many small blockages, and many areas affected by ischaemia, build up and something which may be unnoticeable day to day becomes obvious over the years. There may be a gradual decline in memory and slip into dementia, as more and more tiny areas of the brain have their blood supply interrupted. This is called 'multi-infarct dementia'.

The causes of infarction in the brain are the same as those in the heart: narrowed arteries and a tendency to clotting. It won't come as a surprise that the ways to prevent it, or at least to reduce the risk of it happening, are the same. Stopping smoking, taking measures to lower high blood pressure, being more physically active, losing weight and controlling diabetes all lower your risk of stroke. The 'bad fats' mentioned in The Stomach and Intestines chapter – the saturated animal fats – are bad for the brain as well as the heart. Omega-3 fatty acids may be good for the brain (but probably not as miraculous as the media are currently claiming!) and eating a healthy diet with lots of fruit and vegetables is good for all organs.

Alzheimer's disease

Dementia is a progressive loss of brain function, usually affecting the whole brain, with intellectual, emotional, social and behavioural decline. It is very common, affecting ten per cent of people over the age of 65 and 20 per cent over the age of 80. Alzheimer's disease is the most common form of dementia but 'loss of mind' can also be caused by vascular disease – the multi-infarct dementia I mentioned above.

Alzheimer's disease is characterized by memory loss, problems with language, the inability to recognize things and people, and clumsiness. Someone with Alzheimer's disease may find it difficult to plan ahead, they may become agitated or aggressive, and tend to wander off. Some sufferers

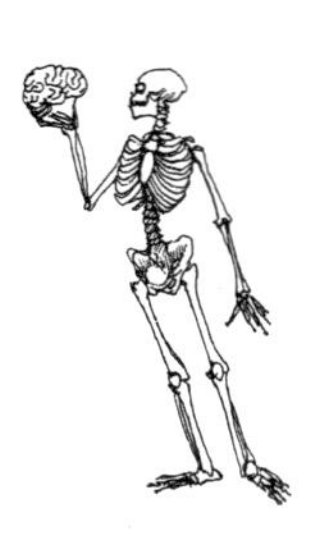

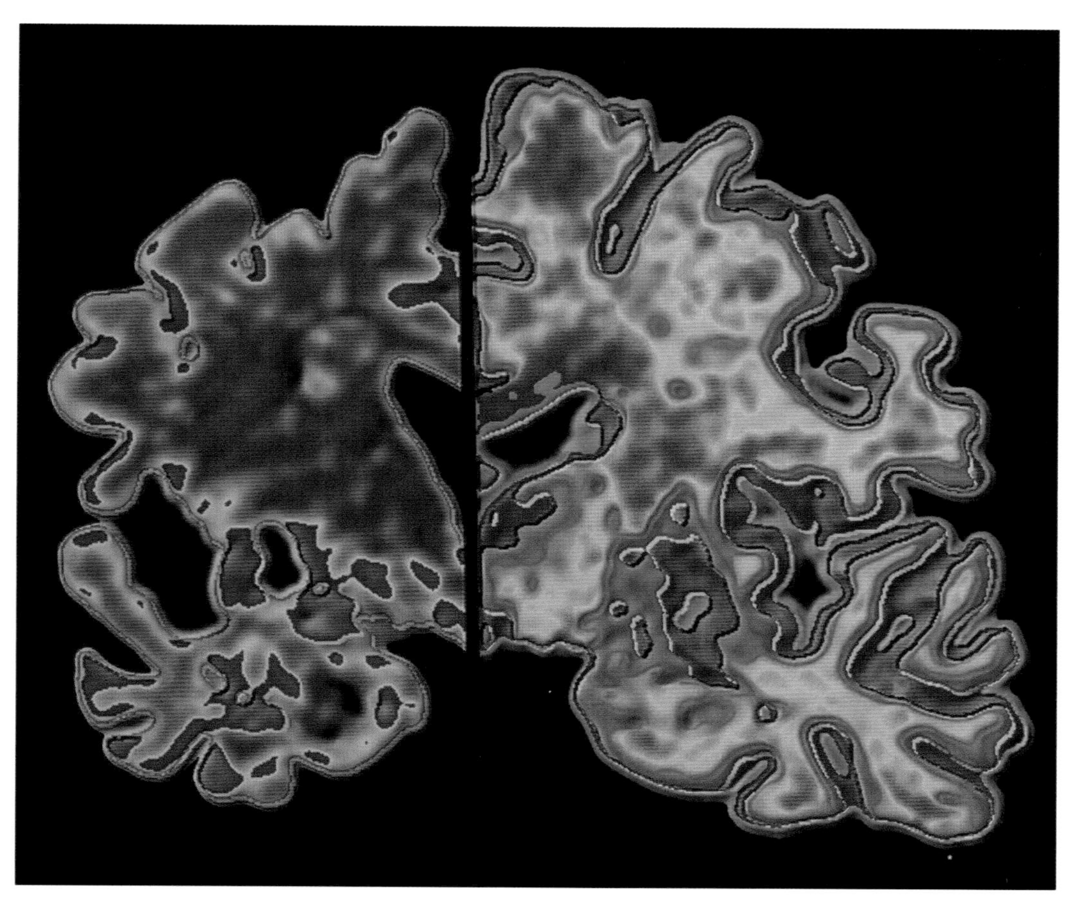

A slice through the brain showing an Alzheimer's patient's brain on the left and a normal brain on the right.

are unaware of the changes happening to them (and this can be very difficult for their friends and family to cope with) but some become very depressed. It is a progressive disease; its effects worsen over time until, eventually, the sufferer dies.

There are tell-tale signs in the brain of someone who has Alzheimer's disease: tangled proteins inside neurons and lumps of protein outside neurons. Its cause is unknown and seems to involve many factors. Some of your risk of dementia is down to your genes. If you have close relatives who have had Alzheimer's disease, you have an increased risk. But lifestyle factors are also important. People who are physically active for most of their

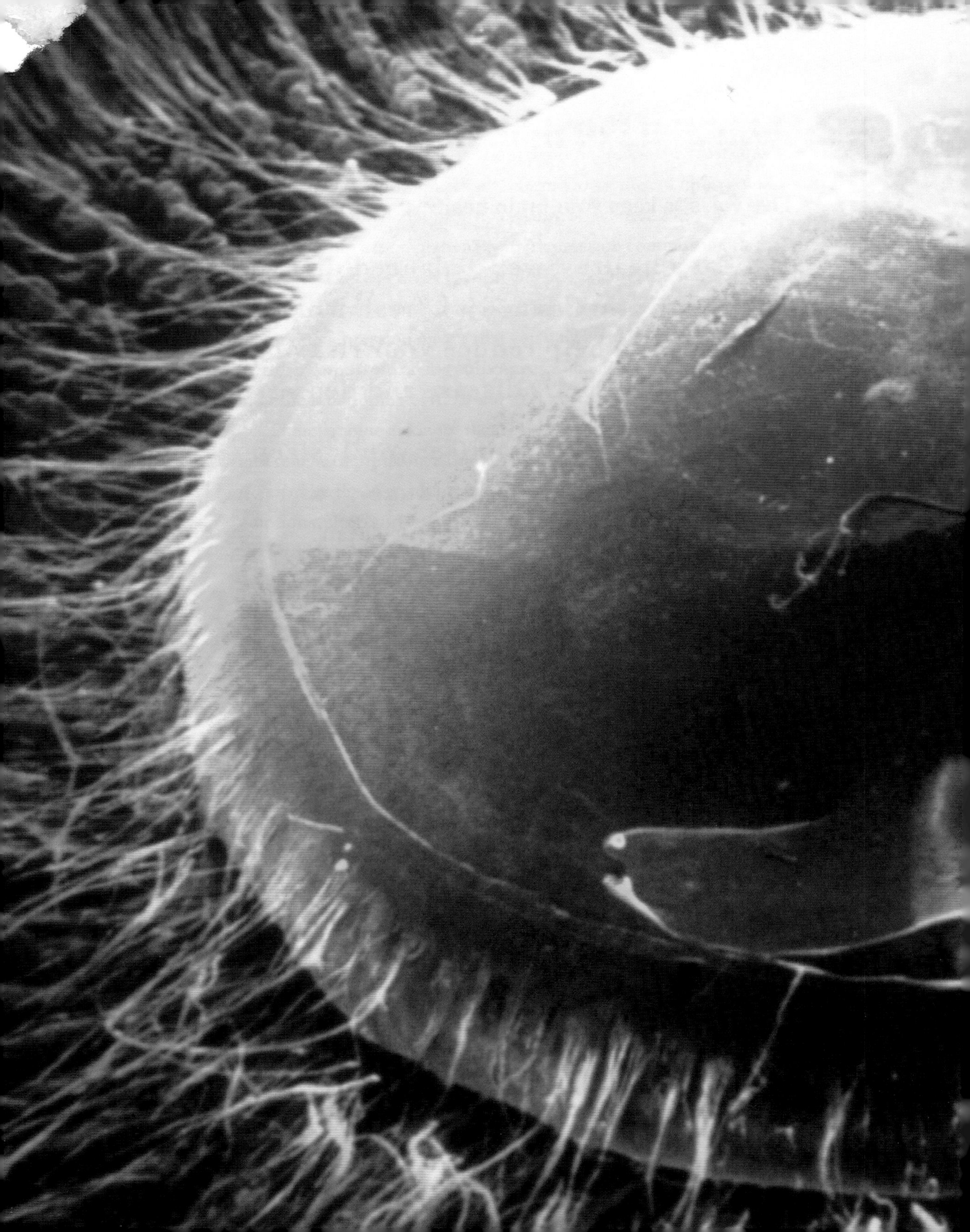

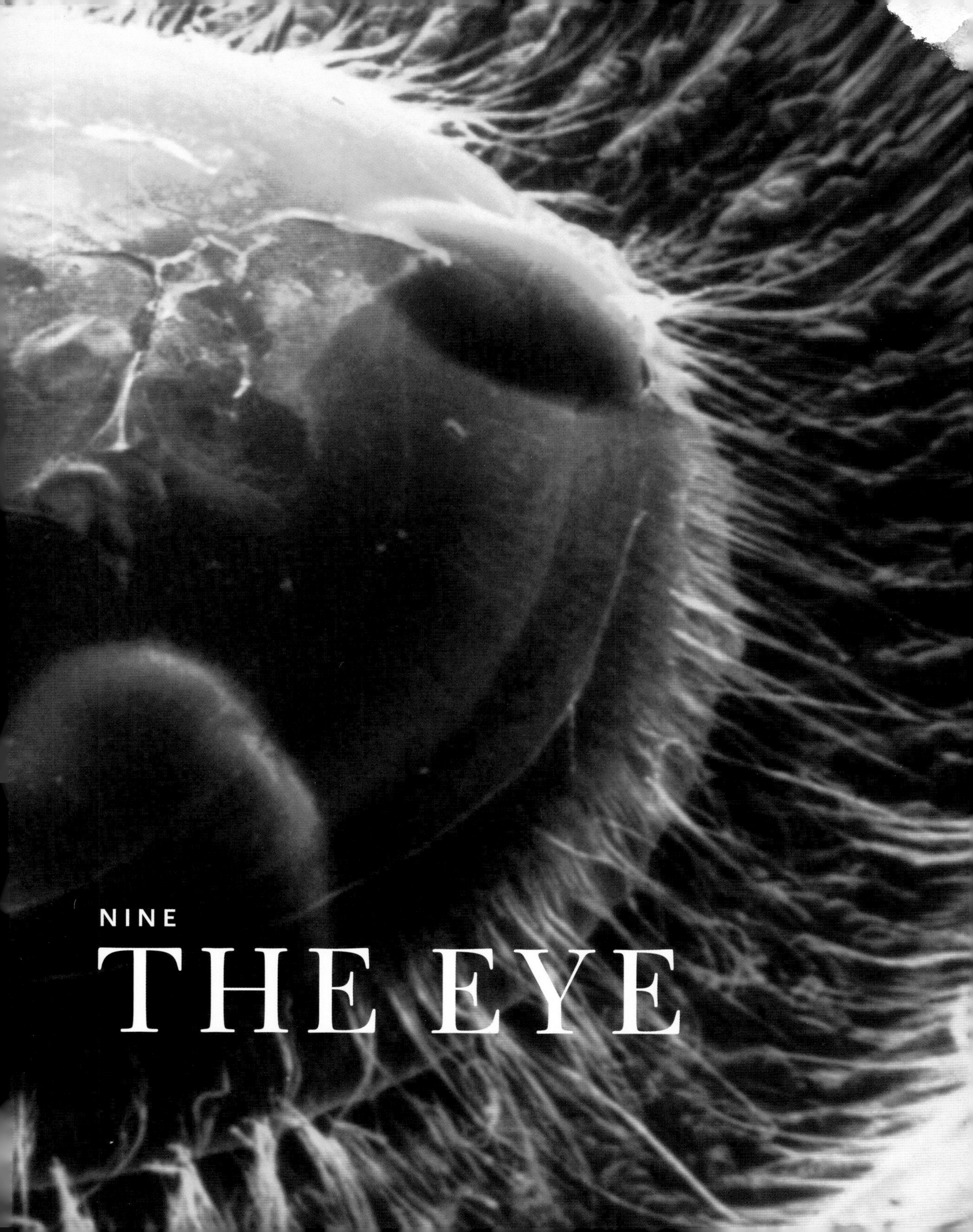

NINE

THE EYE

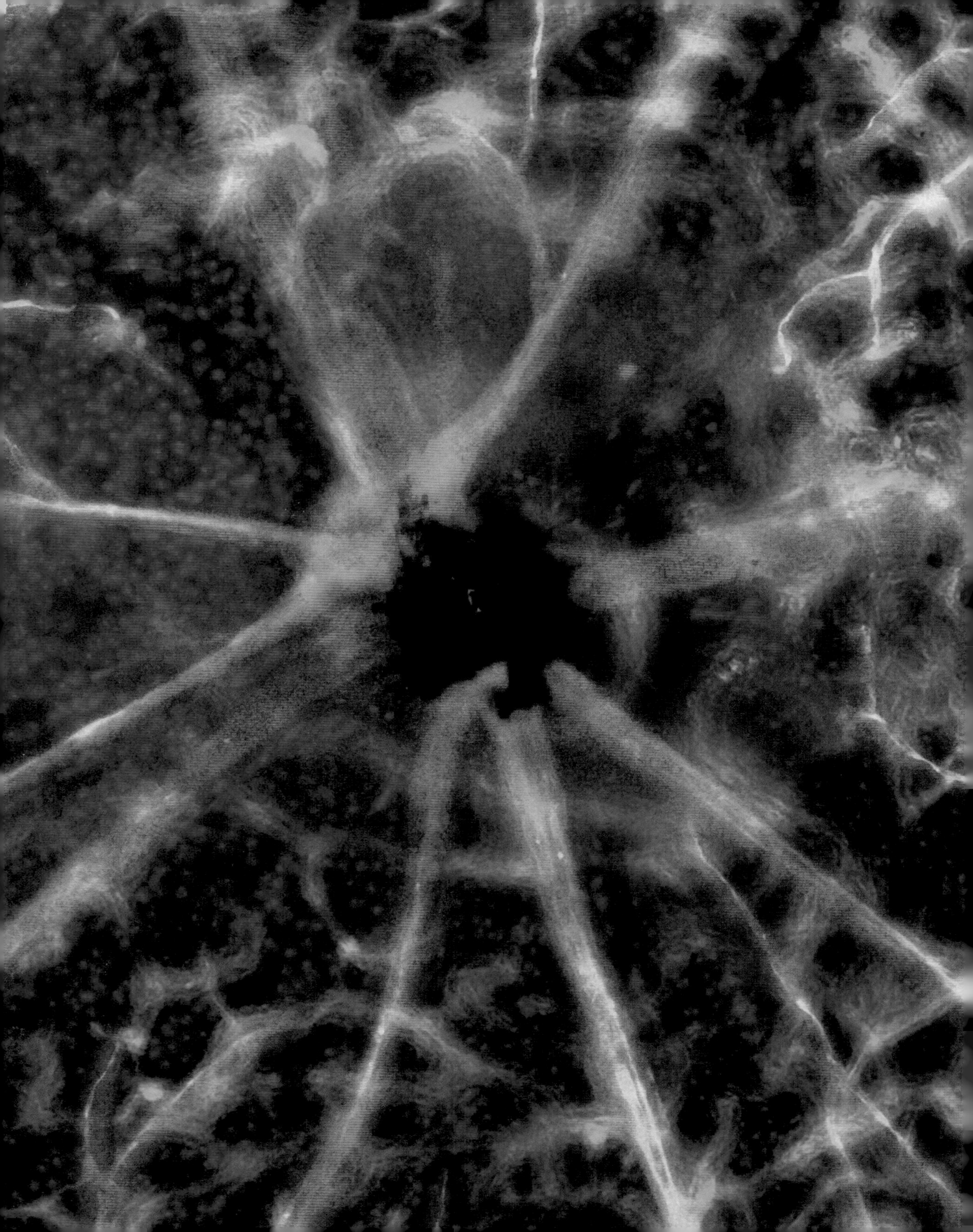

The eyes really are the windows of the soul. They are actually part of the brain: in the developing embryo, the eyes grow out of the brain on stalks, like a snail's eyes. When you're an adult, a doctor can look into your eyes and check on the pressure of the fluid around your brain.

The human eyeball is quite small, on average about 24 millimetres in diameter – smaller than a ping pong ball (38 millimetres). Unlike a ping pong ball, which is full of air, the eyeball is full of fluid. The eyes are an incredible part of the human body. They receive light and generate electrical impulses in nerves, then the impulses travel to the brain, which uses the information to produce a picture of the outside world.

Previous pages **Life through a lens: a circular muscle attaches to the lens via fibres, pulling it into shape and allowing us to shift focus.**

Left **The blind spot: nerves leave, and vessels (red) enter through the same point at the back of the eye. There are no light receptors here, but the brain fills in the missing part of the image.**

How the eye works

My first encounter with eyeballs came in a particularly memorable biology practical class, when the teacher produced a bag of gently defrosting bulls' eyes, straight from the butchers. There was a strange paraphernalia accompanying this early dissection class: oblong enamelled tin bowls filled with a centimetre or so of black wax. I took my bull's eye, placed it in the bowl, from where it gazed up at me dolefully, took up a scalpel and considered my first incision. Twenty minutes later, the eye was open, its leathery coat cut and the vitreous humour (Latin for 'glassy fluid') oozing out. I cut the lens free from the circular ciliary muscle attaching it to its rim

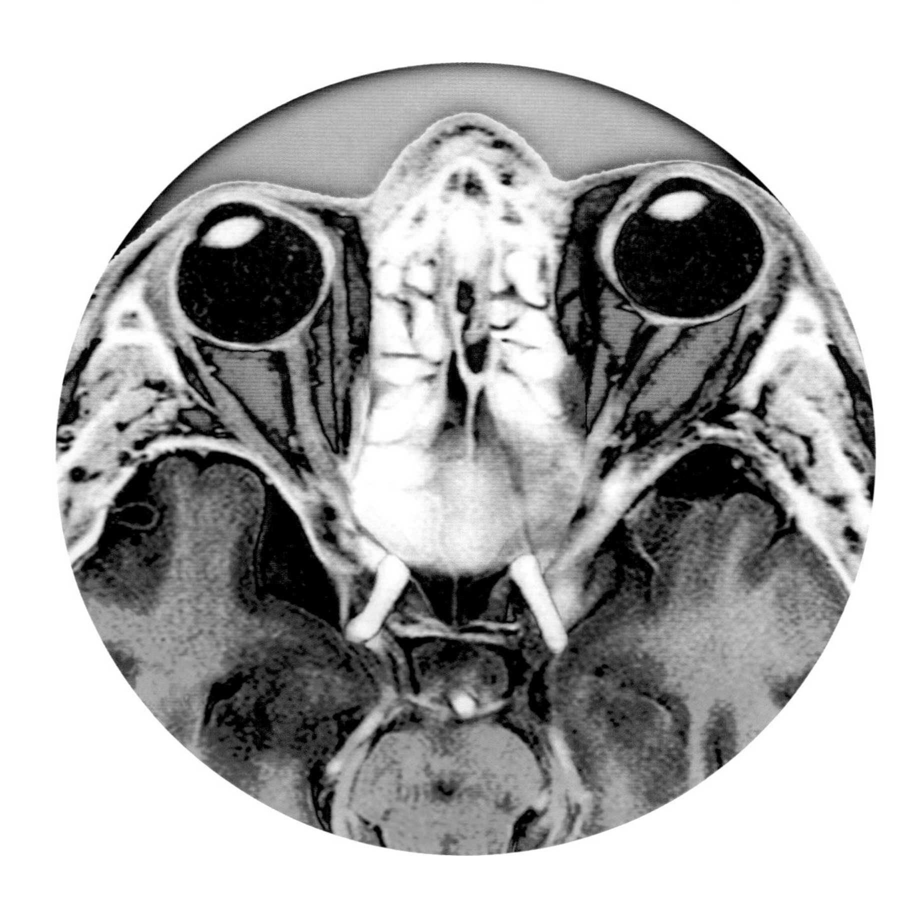

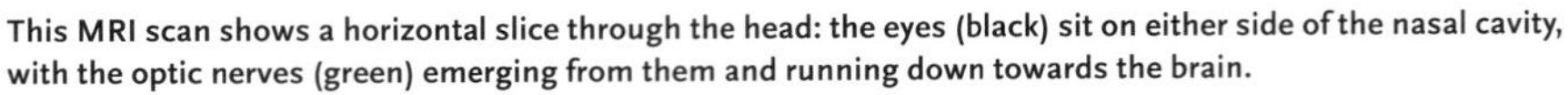

This MRI scan shows a horizontal slice through the head: the eyes (black) sit on either side of the nasal cavity, with the optic nerves (green) emerging from them and running down towards the brain.

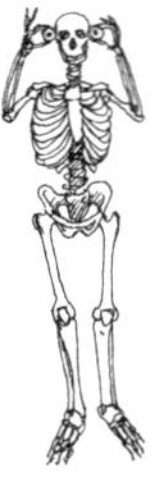

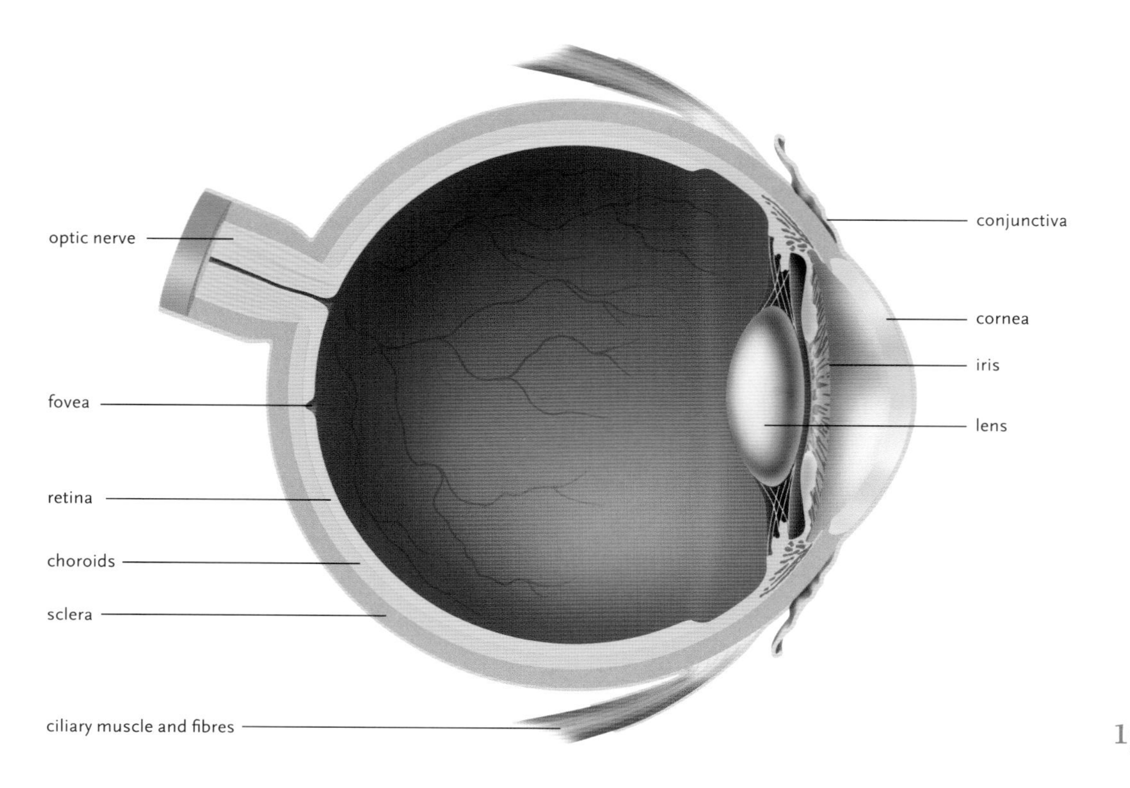

A living camera: the parts of the eye.

and set it down on a sheet of writing, amazed at the magnifying capability of this small lump of jelly. Wonder quickly replaced disgust, although I think a few of my class may have eschewed meat for dinner that evening.

The anatomy of the eye is very like the structure of a camera: the cornea is the lens on the front of the eye-camera, with an additional lens behind it to adjust the focus; the iris controls the aperture; the pupil controls the amount of light entering the eye; and the retina is the light-sensitive film (or sensor in a digital camera). For the eye to work, the cornea, lens and the fluids inside the eye must be kept clear, to allow light through; the cornea must be kept wet, to preserve the smoothness of its surface; the lens must be able to change shape readily, to focus on objects near and far away; the cells of the retina must be healthy and receptive to the light falling on them.

Parts of the eye

The cornea

Most of the eye is covered in a tough, white coat, the sclera, but one fifth of its area, at the front, forms a transparent dome: the cornea. This bends the light entering the eye, providing about eight-tenths of the focusing power of the eye. The cornea and the sclera around it are covered in a membrane that turns back on itself to line the inside of the eyelid – the conjunctiva. The conjunctiva is constantly bathed in a thin film of tears, which cleans the surface of the eye and keeps the surface of the cornea smooth so that light is bent, or refracted, evenly. Blinking your eyelids replenishes the tear film and stops the cornea drying out. Tears are produced by the lacrimal gland, which sits on the upper, outer side of each eyeball and releases its secretion under the eyelid. The tears move across the eye to the inner corner, where the fluid disappears down two tiny holes and into two tiny canals, which empty down the nasolacrimal duct into the bottom of the nose. This is why, when you cry, your nose runs.

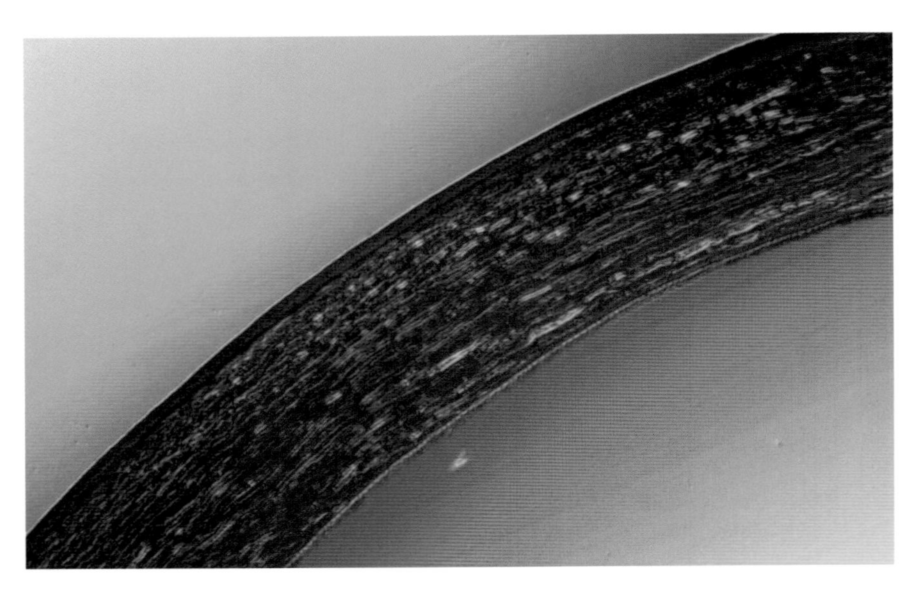

The transparent, dome-shaped cornea of the eye is seen under the microscope. It has no blood vessels and gets its oxygen from the air and its nutrients from the fluid inside the eye.

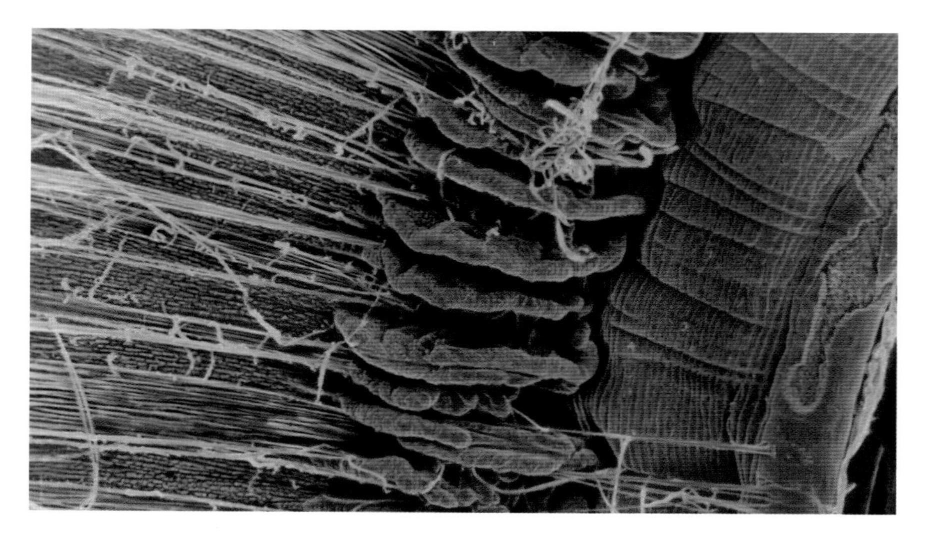

A rainbow: the pupil is the dark area in the far bottom right; the muscles and fibres of the iris control the size of the pupil.

The iris

The iris controls the amount of light entering the eye and also controls the direction of rays coming in, by widening or shrinking the pupil. The iris has both circular muscle fibres (around the pupil) and radial muscle fibres (radiating out like spokes on a bicycle wheel). The circular fibres close the pupil down, and the radial fibres pull it wider: the iris muscle can change the diameter of the pupil from one millimetre to eight millimetres.

The pupils are large in darkness and small in bright light. A wide-open pupil will admit rays in many directions, whilst a closed pupil lets in parallel rays of light. This alters the depth of focus just like the aperture on a camera, and when you look at something near to you, the pupil constricts to sharpen the focus. If you have an eye test in hospital, the doctor will usually ask you to focus on something far away, and then something near, to check that your pupil reacts and gets smaller. It's not actually your pupil that reacts, because the pupil is a hole; it's the iris muscle. The iris (Greek for 'rainbow') comes in a selection of colours, from pale blue, through green, to dark brown. The pigment in the iris (melanin) is the same as in skin: the more pigment in the iris, the browner it is. There is less pigment in babies' eyes – they tend to be blue at birth, even if they later turn brown.

The lens

The lens is supported around its edges by the circular ciliary muscle and ligament. As the muscle contracts, it pulls inwards, reducing the diameter of the lens and allowing it to bulge out; this is what happens when you focus on something near, and the fat lens bends the light rays to focus them on the retina. When you focus on a distant object, the ciliary muscle relaxes back out, and the lens is pulled into a thinner shape, hardly bending the incoming light at all. It's incredibly clever: rather than having to have a set of lenses to focus on objects at different distances away from you, just one adjustable lens does the job. The lens gets tougher as you get older and doesn't change shape as efficiently any more, tending to settle into short-sightedness.

The lens isn't a bag of clear fluid. Remarkably, it's a lump of cells: a tissue. These long, crystal-shaped cells work hard to pump out anything which might turn them cloudy. With age, the lens cells tire of pumping out clouding particles and the lens can turn milky: this is what a cataract is. I'm not sure why this word has come to be used for this particular cause of blindness. Perhaps it's because slow flowing streams are clear but the water turns white in a cataract.

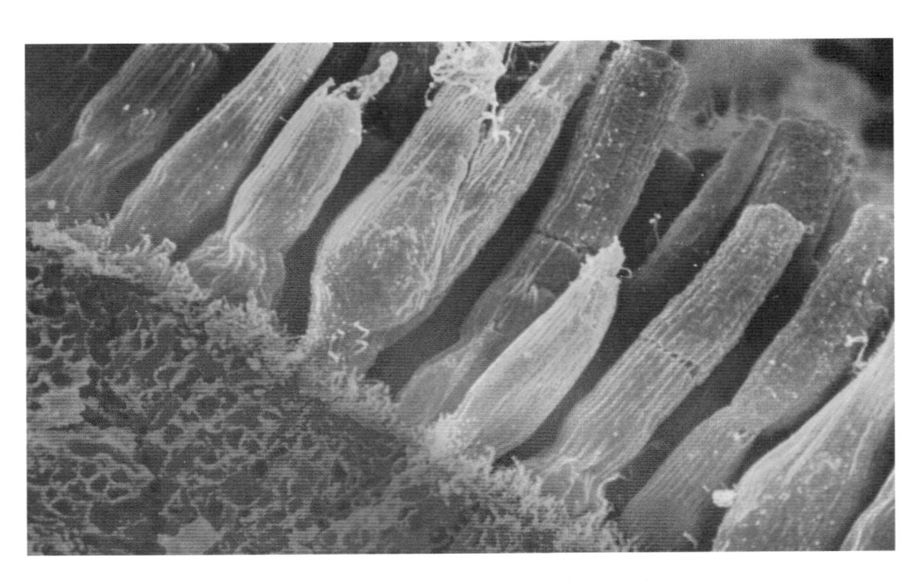

There are about 130 million rod cells and 6.5 million cone cells in the human retina. Magnified nearly 8,000 times, the rods here are coloured blue and the cones, green.

The retina

Inside the tough outer coat of the eye (the sclera), is the choroid, a layer replete with blood vessels. Inside the choroid is the retina, where the real business of the eye goes on.

In the retina, light is transformed into electricity. The retina is about a tenth of a millimetre thick and has two layers: a layer of light sensitive cells (photoreceptors) spread over a layer of pigment cells. The photoreceptors catch the light entering the eye and the pigment layer acts like a dark curtain, quenching the light and preventing it from bouncing around inside the eye and producing multiple images.

There are two types of photoreceptors: rods and cones. The cones register colour but need plenty of light, whereas the rods don't register colours but can work in dim light. Our colour vision is part of our primate heritage; many mammals don't see the range of colours that we do but, thanks to our ancestors who lived in trees and needed to be able to spot fruit amongst green leaves, we see the world in glorious colour.

Looking at the retina with an ophthalmoscope, the area where the optic nerve leaves the eye, the optic disc, looks like a pale doughnut. The retinal artery enters the eye in the middle of the optic disc, then branches to supply the retina with oxygenated blood. There are no photoreceptors in the optic disc, so it's known as the 'blind spot'. You don't notice it normally, because your brain very cleverly fills in the hole – but you can find it: take a pencil and hold it at arm's length, tip pointing up. Close one eye and stare right ahead with the other one – don't move it at all. Move the pencil around slowly. If you do it properly, keeping your open eye still and not letting it move to look at the pencil, you should notice a point – slightly to the outer side of the middle of your field of vision – where the pencil tip disappears. You've discovered your blind spot.

When you look at the retina, you can also see an area that's a different colour from the rest; this is the macula lutea (Latin for 'yellow spot'), where there is a particularly dense cluster of cone cells. At its centre is a small pit, the fovea. The fovea is the part of the eye that gives us the most detailed, fine resolution picture of the world we look at.

Common eye problems, and how to prevent them

The eyes are tender, vulnerable parts of our anatomy. They have eyelids to help protect them from physical insult, but one of the things that causes long-term damage to the eyes is invisible: ultraviolet light.

Macular degeneration

As we get older, the macula grows older too. Because this cluster of cells is where we see with the finest resolution, the detailed area in the centre of our vision is gradually lost. Macular degeneration is the most common cause of blindness in the developed world – more common than cataracts. And as more of us are surviving into old age it can only increase.

The precise way in which macular degeneration happens isn't properly understood. It seems to be something to do with cells in the retina getting old and tired, and waste products building up around them (it's like they've forgotten to put the bins out). Eventually the cells give up completely: vision in the middle of what we look at blurs, then completely disappears.

Although the way the disease develops may not be entirely clear yet, the degeneration is certainly linked to age, smoking and obesity. You may not be able to do anything about the first one but you certainly can about the other two. Light is also a factor: it's paradoxical that the very stuff the eye needs to produce vision can also be damaging.

As well as the more well-known dangers of ultraviolet radiation, it seems that the eye can also be damaged by visible blue light. Light damages the retina in several ways: by creating small shock waves which can cause irreparable mechanical damage to both photoreceptors and the pigment epithelium beneath, by raising the temperature of the retina; or by creating free radicals, which attack all sorts of molecules, often creating a domino effect. We all know that it's dangerous to look at the sun as it can damage the retina. (In 1912, thousands of people in Germany suffered macular damage from viewing a solar eclipse.) In the long term, exposure to intense sunlight – even without looking at the sun directly – increases the risk of macular

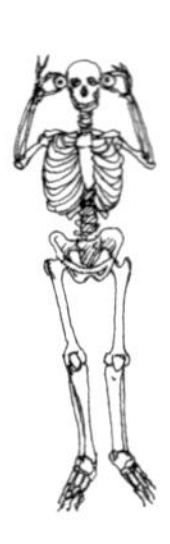

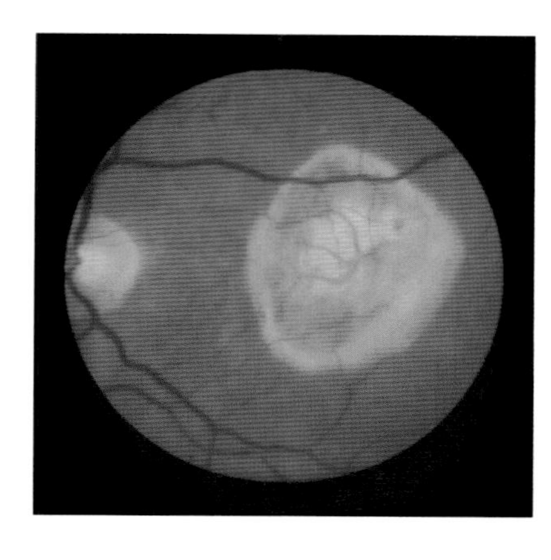

Damaged retina: looking through an ophthalmoscope, the optic disc is the yellow circle on the left; the yellow area on the right is macular degeneration.

degeneration. So get those sunglasses on.

Some studies have suggested that high doses of antioxidant vitamins (like A, C and E) can reduce the risk of macular degeneration – but high doses of these vitamins can be bad for your health. There is compelling evidence that carotenoids (like lutein and the exotic-sounding zeaxanthin) help to slow down the visual deterioration and perhaps even produce an improvement in vision, but more trials are needed. Lutein (from the Latin for 'yellow') is a yellow pigment found in egg yolk, and both it and zeaxanthin (Greek for 'maize-yellow') are found in high levels in green, leafy plants. Even though our own bodies don't manufacture these carotenoids, they're nonetheless present in a variety of tissues: in our skin, blood and eyes and, guess what, in particularly high concentrations in the macula. It seems sensible that eating high levels of these substances might protect against macular degeneration: they are both antioxidants and because they are pigments, might also act like a 'sunscreen', protecting the retina from blue light and ultraviolet radiation.

If you want to try to maximize your intake of lutein and zeaxanthin, leafy green vegetables are packed full of the stuff. Raw kale (though probably not the most appetizing!) tops the list, with a whopping 39 milligrams per 100 grams, followed, some way behind, by other green things such as spinach, peas, Brussels sprouts, broccoli and beans. Ophthalmologists generally advise people who have macular degeneration to stop smoking and to eat lots of fresh fruit and vegetables. This is equally good advice for people who want to reduce their risk of developing macular degeneration.

But do beware of miracle cures. Because macular degeneration is one of those conditions where treatment isn't very effective, there are a lot of people claiming that their supplements will work.

Cataracts

Cataract surgery is the most common operation in the UK: a quarter of a million cataract operations are performed every year and it's a very effective surgical treatment where the original clouded lens is removed and replaced with an artificial lens. But is there anything you can do to reduce the risk of getting cataracts? Certainly, there is a genetic component to cataracts and age is also a factor; neither of which you can do much about. Ultraviolet radiation contributes to cataract formation and may account for the high numbers of cataracts in sunny countries like India. Like I said – get those sunglasses on. Diabetes is another risk factor. So anything that reduces your risk of developing diabetes – such as watching your weight and keeping physically active – will help reduce your risk of developing a cataract.

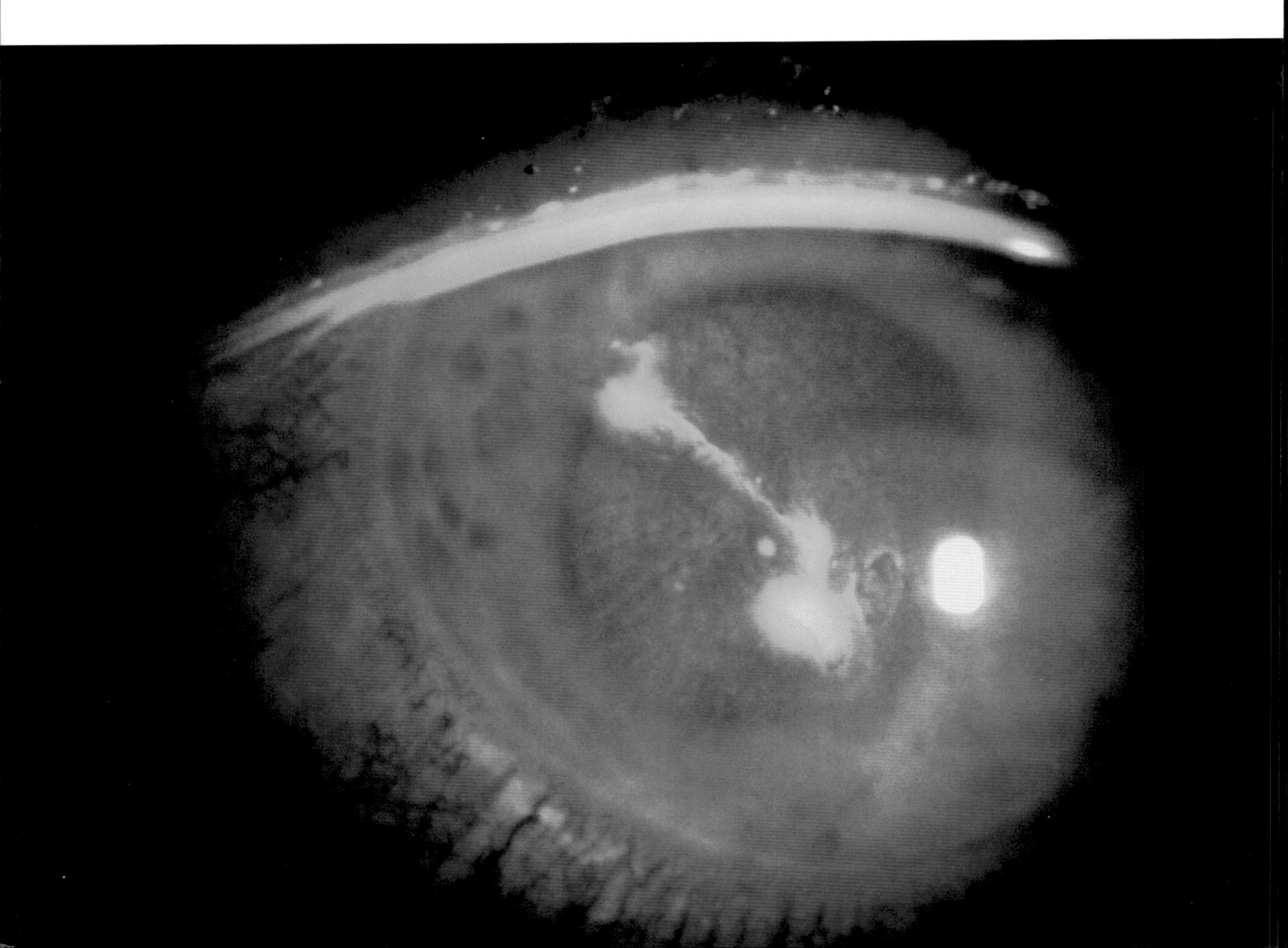

Corneal abrasions

Although the cornea may be tough, it's not invincible. It's important to protect your eyes whenever you're doing anything that might cause bits to fly up into them. Scratches of the cornea are very painful, making you blink a lot and causing your eyes to water profusely. Corneal abrasions make infections more likely – so it's important to see (with your good eye) a doctor who will prescribe antibiotic eye ointment or drops.

Conjunctivitis

Inflammation of the conjunctiva, which lines the surface of the eye, can turn the white of the eye red as the blood vessels enlarge. This inflammation might be due to an allergic reaction, or a bacterial or viral infection.

You can help control allergic conjunctivitis by avoiding the particular allergens that set you off. If pollen is the allergen, try to stay indoors with doors and windows closed when the pollen count is high. If it's fungal spores or house dust mites, frequent vacuuming and dusting (sorry) can cut down your exposure. And when it all gets too much, a cold, wet flannel over the eyes can be very soothing.

Viral conjunctivitis, which is more common than bacterial, might appear after a cold or a flu-like illness. It's extremely important to stick to a strict hygiene regimen, as it's extremely infectious. Wash your hands frequently, avoid sharing flannels, towels and pillowcases and wash these regularly. Don't share cosmetics like eyeliner and replace any cosmetics you used while you had the infection or you'll find yourself re-infected. Although viral conjunctivitis usually clears up by itself, it's important to see your doctor to rule out other possible infections.

Eye strain

Focusing on something very close to you for a long period of time – whether a book, a computer screen or any other close work – can cause eye strain. Symptoms of eye strain include a feeling of tightness around the eyes, difficulty changing focus and temporarily blurred vision. Eye strain is often accompanied by a headache, which may range from a dull ache to sharp pain.

Scratches on the cornea: with fluorescein drops in the eye, under ultraviolet light, a corneal abrasion can be clearly seen as a green track across the pupil.

As more and more of us use computers in our jobs, eye strain related to computer use is becoming much more common. The symptoms of eye strain: tired eyes, irritation, redness, blurred vision and double vision are sometimes bundled together in a condition called 'computer vision syndrome'. Twenty years ago, before computers had begun to take over the workplace completely, office work tended to involve more activities and more moving around, producing natural breaks from sitting at a desk. Many people now have computers at home as well, so the problems aren't restricted to the workplace.

Initial research into the potential harms of using computer screens focused on radiation from the screen but didn't really find any negative effects. The strain we put our eyes under when focusing on a screen is the real problem. The most common computer-related symptoms are eye problems affecting the surface of the eye, the focusing apparatus inside the eye and the muscles that move the eye. Dry eyes are caused by insufficient blinking. If we forget to blink – which it seems most of us do when we stare at a computer screen (like I am now, typing this book!), the surface of our eyes dries out. This has a few consequences: it might make our eyes feel gritty and blurred, and our eyes water to compensate. The drying of the surface of the eye might be made worse by looking at a bright screen in a darkened room, and by air conditioning. Whereas we might look down to read a book, with our eyelids half-covering our eyes, computer screens are usually straight ahead, so more of the surface of the eye is exposed when we look at them. We produce fewer tears as we age, so older people are even more likely to get dry eyes, especially post-menopausal women. If you're wearing contact lenses and you have dry eyes, then the lenses can stick to the eyelids, causing extra discomfort.

Looking at a piece of white paper with dark letters on and frequently swapping to look at a bright screen with a dark background tires the iris muscle, so it's best to stick to dark characters on a light background on the screen. You should avoid too strong a contrast between the brightness of the computer screen and the general brightness of the room, so keep the light levels up in the room around you but, at the same time, try to avoid glare on

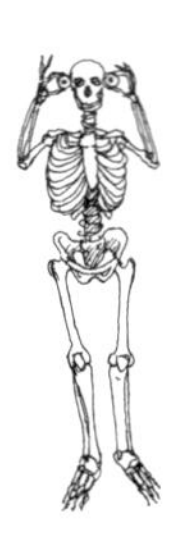

Look into my eyes: an optician can check acuity of eyesight as well as the health of the eyes, but this Victorian optician seems to be keeping his distance!

the screen. Some studies have shown that screen filters, which cut down glare, do help relieve the symptoms of computer vision syndrome, but more work needs to be done. Flat screen (LCD, including TFT) monitors are pretty much flicker-free and less likely to cause eye strain than the old cathode ray tubes (the ones that look like old televisions).

Doing a lot of close work affects the focusing apparatus of the eye, causing transient short-sightedness in some people. However, there's no evidence as yet to suggest that this temporary short-sightedness will increase your risk of permanent short-sightedness. If you're struggling to focus on the computer screen, then the twenty-first-century version of reading glasses – computer glasses – can help relieve eye strain.

Tips for people regularly using computer monitors – that's more and more of us these days – include making sure that your position at the desk is comfortable, to avoid neck and back strain. You should try to position your monitor so that it's as far away as you can manage, but so that you can still comfortably see the screen. Tilt it to about 10 to 20 degrees, so that the top of the screen is slightly further away than the bottom. Try to take regular breaks, and actually get up and walk around rather than just surfing the web! (Note to managers: studies show that breaks improve efficiency – people make up for the time 'lost' in the break.) Tear substitutes may also be useful if you get dry eyes and the measures above haven't solved the problem.

Last – but definitely not least – make sure you have regular eye tests. Eye tests are definitely A Good Thing. The eye is a useful 'window' on your health: the optician can check the health of your eyes as well as looking out for any early signs of conditions including diabetes and high blood pressure by examining your retina.

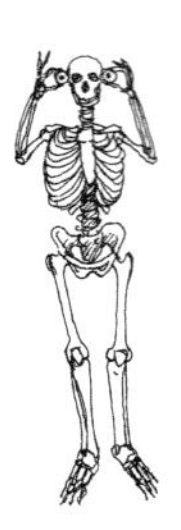

Five ways to keep your eyes healthy:

- Have your eyes tested regularly to check your sight, the health of your eyes, and for early signs of other health problems such as high blood pressure and diabetes.

- Protect your eyes from intense sunlight to reduce your risk of macular degeneration and cataracts.

- Protect your eyes from abrasions: wear glasses, goggles or visors if there's any chance of something getting in your eyes.

- Eat a healthy diet, with plenty of fruit and vegetables, to reduce your risk of macular degeneration and cataracts – and don't smoke.

- Take frequent breaks when working on a computer, to avoid eye strain.

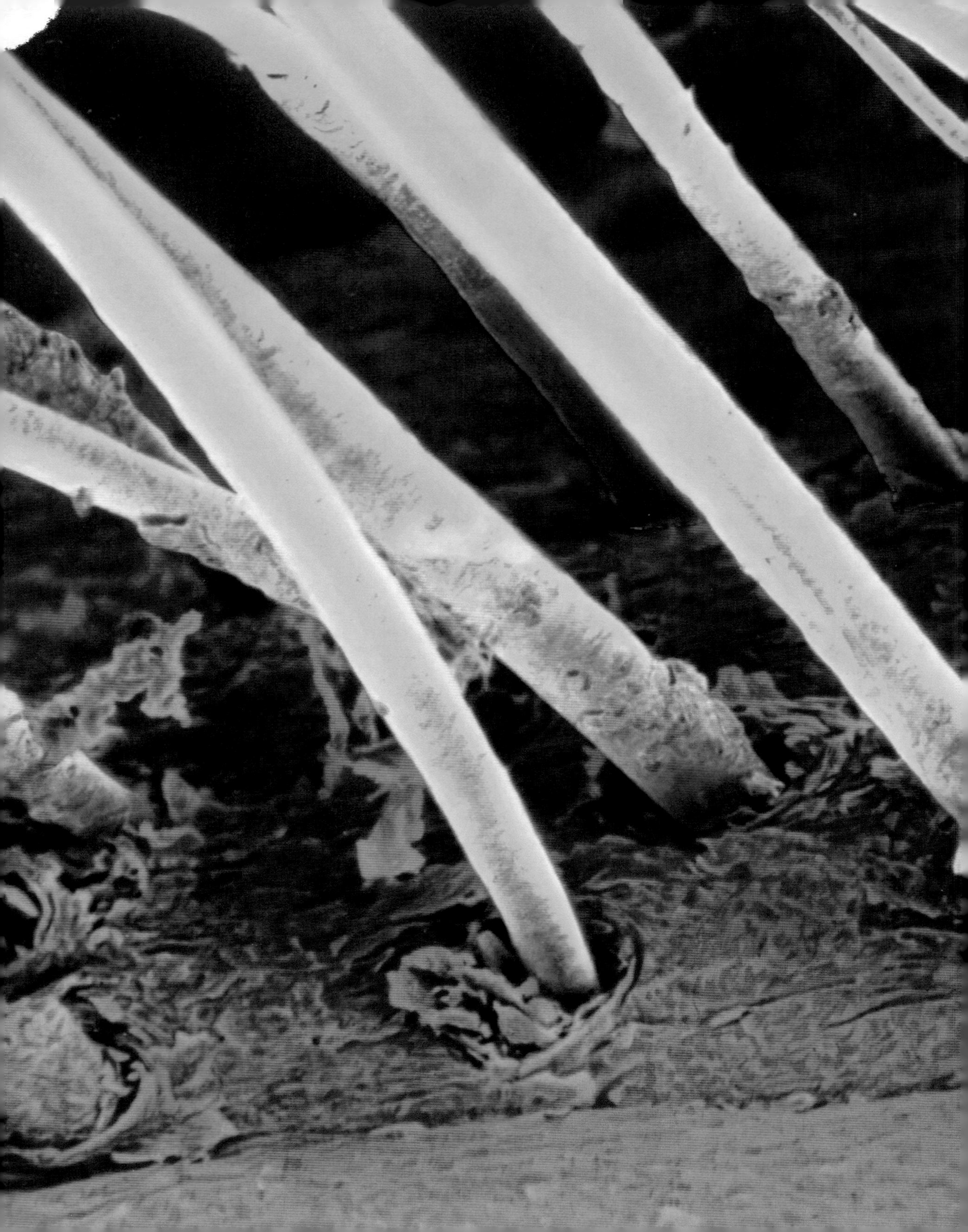

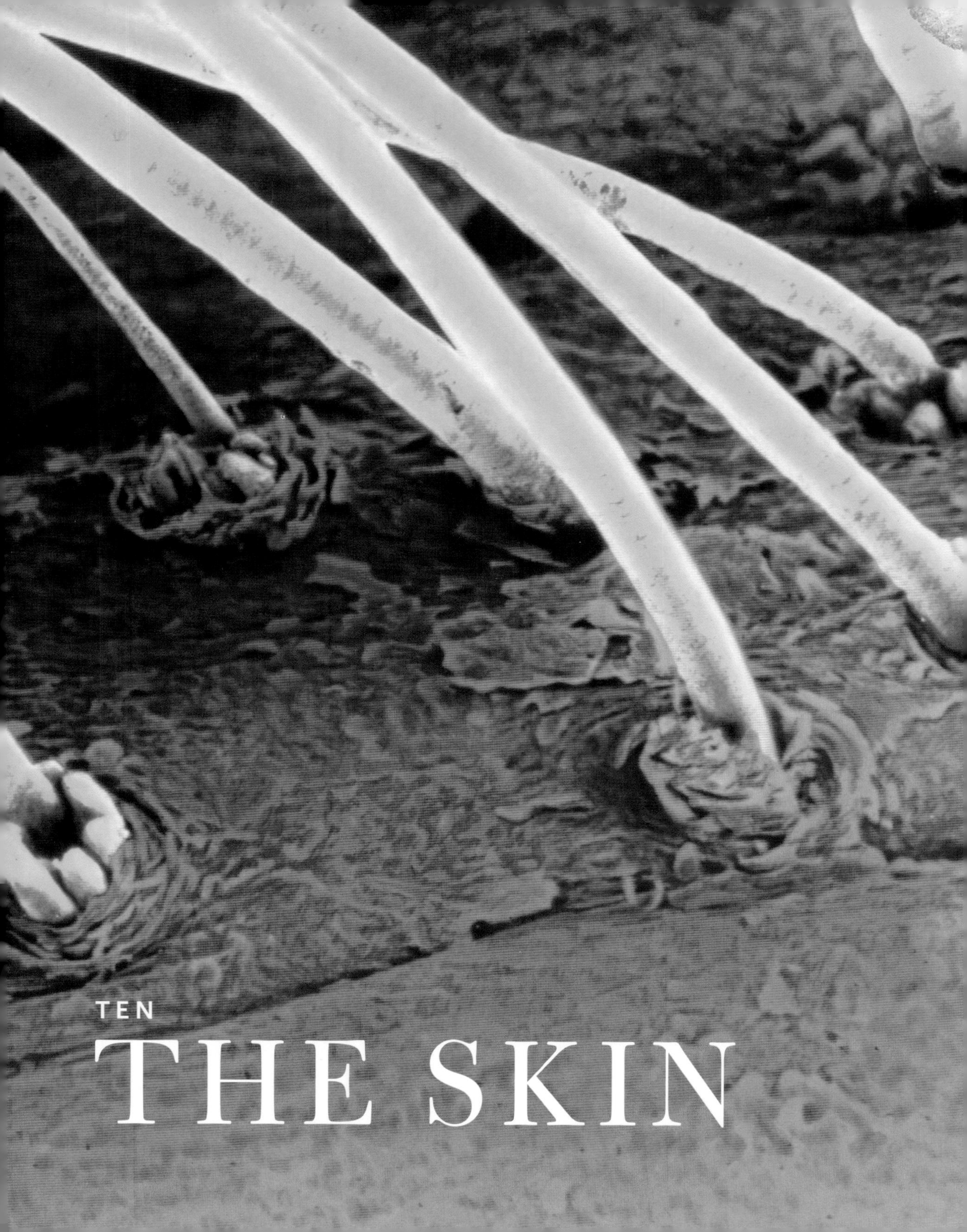

TEN

THE SKIN

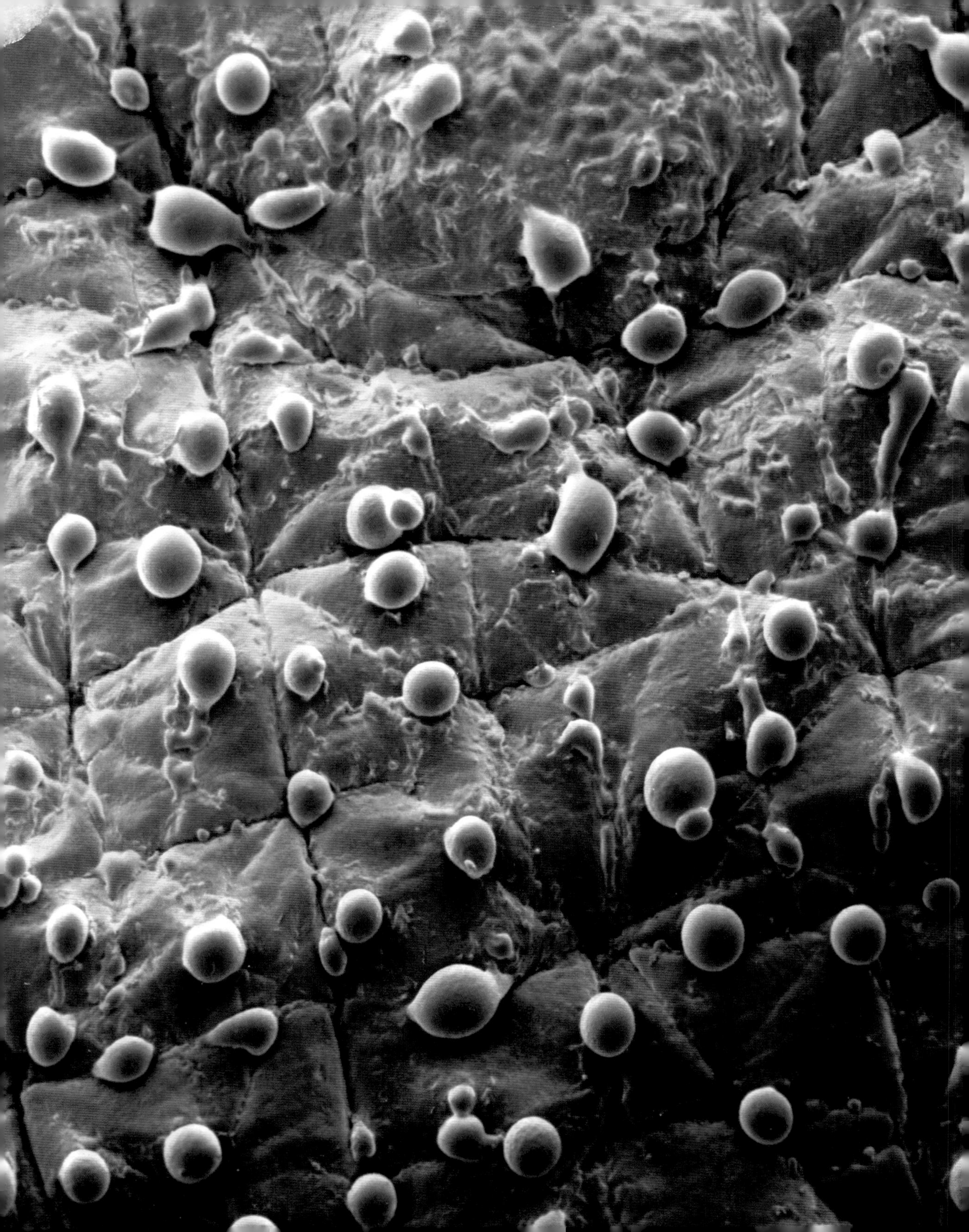

The other organs we have met have all been on the inside. I'll finish my tour by looking at the organ that forms our outside.

The skin is an amazing organ. It weighs about five kilograms and covers an area of about two square metres. Every square centimetre of skin contains, on average, more than 200 sensory receptors, 15 sebaceous glands, 100 sweat glands, 55 centimetres of nerves and 70 centimetres of blood vessels.

It is the organ through which we sense the texture and temperature of our environment; it regulates the temperature of our bodies; it makes vitamin D; it excretes waste as sweat; on our fingers, its ridges allow us to grip things; and it communicates that we're embarrassed. The skin is our waterproof outer lining, protecting us from our environment and keeping the outside out and the inside in. We have to be careful not to take it for granted, but to look after it and protect it, if we are to benefit from its protection.

Previous pages **The eyelids are fringed with hairs to stop dust getting into the eyes.**

Left **Beads of sweat: blue droplets emerge from sweat pores on the skin of the back of the hand.**

The layers of the skin

The skin has two layers: the outer epidermis and the inner dermis. The epidermis is a tough, waterproof covering, built up of layers of cells. Cells are constantly sloughed off from the surface, but replaced from beneath: the epidermis constantly renews itself. Under the epidermis is a layer of connective tissue called the dermis, containing the blood vessels and nerves of the skin. Under the skin is the subcutaneous tissue, which, in most areas, contains numerous fat cells.

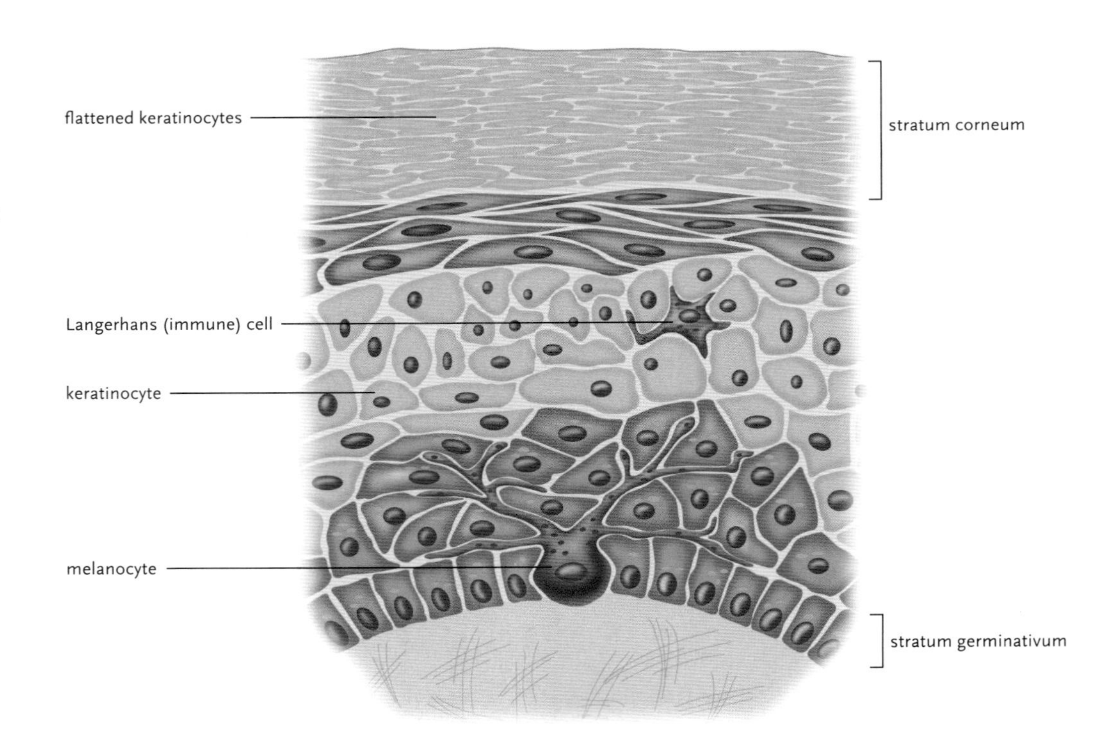

Multi-storey protection: layers of skin cells cover the outside of our bodies.

The epidermis

The thickness of the epidermis varies from the very thin epidermis on your eyelids (about 0.1 millimetres thick) to the thick layer on the palms of your hands and the soles of your feet (about one millimetre thick). But even in the thinnest epidermis, there are several layers of cells, most of which are the keratin-producing keratinocytes.

The bottom-most layer of the epidermis is the stratum germinativum, which just means 'germinating layer'. It's from this layer that the skin constantly produces new cells. As well as containing keratinocytes, it contains cells of the immune system and the pigment-producing melanocytes. The pigment that these cells produce, melanin (Greek for 'black'), absorbs ultraviolet radiation and protects our DNA. It also scavenges free radicals.

The colour of our skin varies from person to person; each person's skin also changes colour over the course of the year. People from different populations throughout the world have different levels of melanin in their skin, producing all the variation in skin colour from the deepest ebony to an ivory white.

The level of pigment seems to depend on how large and active our melanocytes are; we all have similar numbers of these pigment-producing cells, regardless of the colour of our skin. People with freckles actually have fewer melanocytes in the freckles than in adjacent skin, but the freckle melanocytes are bigger and more active. As well as having a genetically determined level of pigment, there's also room for adaptation, according to how much sunlight we are exposed to: we tan. Tanning happens in two ways: there's an immediate effect, as existing pigment in the skin gets darker, and a delayed effect when the melanocytes start to work harder to produce pigment. Hormones can also cause changes in pigmentation. For example, during pregnancy, the raised levels of oestrogens and progesterone cause darkening of the skin, especially on the face, nipples, abdomen and genitals.

The uppermost, surface layer of the epidermis is the stratum corneum. Skin cells produce keratin, a tough protein that also forms the basis of our nails and hair. The cells in the stratum germinativum are plump but gradually become flatter and flatter until, at the surface, the stratum corneum cells aren't even really cells any more: they're more like scales, or flat packages of keratin. The surface layer of skin is meant to wear away – the surface cells can be rubbed off without really damaging the skin – so the skin constantly renews itself, as cells at the bottom of the pile divide and move up towards the surface. Your skin totally renews itself every month and we each shed around 19 kilograms of dead skin in our lifetime.

The dermis

This deeper layer is made of a strong but flexible connective tissue, with collagen-producing cells (fibroblasts) and various immune system cells sitting in a matrix of collagen and the elastic protein, elastin. Although the dermis is strong and elastic, it can be torn by stretching, which produces silvery striae: stretch marks. If the skin is rubbed or burned, the epidermis can separate from the dermis, forming a blister. The dermis is the part of the skin that contains blood vessels, hair follicles, sweat and sebaceous glands.

The dermal blood vessels of the skin lie in dense networks or 'beds' of capillaries. When you're too hot, small arteries (arterioles) widen to allow more blood to flow into these networks, so that heat from the blood can be lost through the skin, and your skin becomes red and flushed. When you're cold, the little arterioles feeding the capillary beds narrow right down and very little blood flows into the capillaries. The skin becomes white and cold, reducing the amount of body heat that you lose at the skin's surface. As well as an important device for regulating body temperature, the changeable blood flow to the skin can also communicate emotions, as we flush with embarrassment or blanch with fear.

The sweat glands also lie within the dermis: coiled tubes which extend upwards to open at a pore on the surface. They are very important in temperature regulation. Sweat glands are stimulated when the body is hot and pour their salty contents on to the surface of skin. The sweat evaporates, taking away the heat from your skin and producing a cooling effect. Sweat glands are found all over the body (except the lips, nipples and parts of the external genitalia). Sweat glands in the armpits and genital areas open into

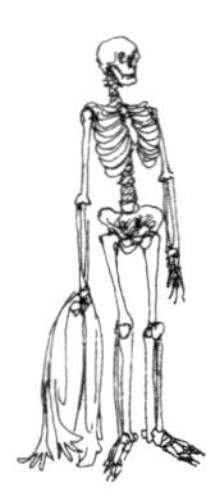

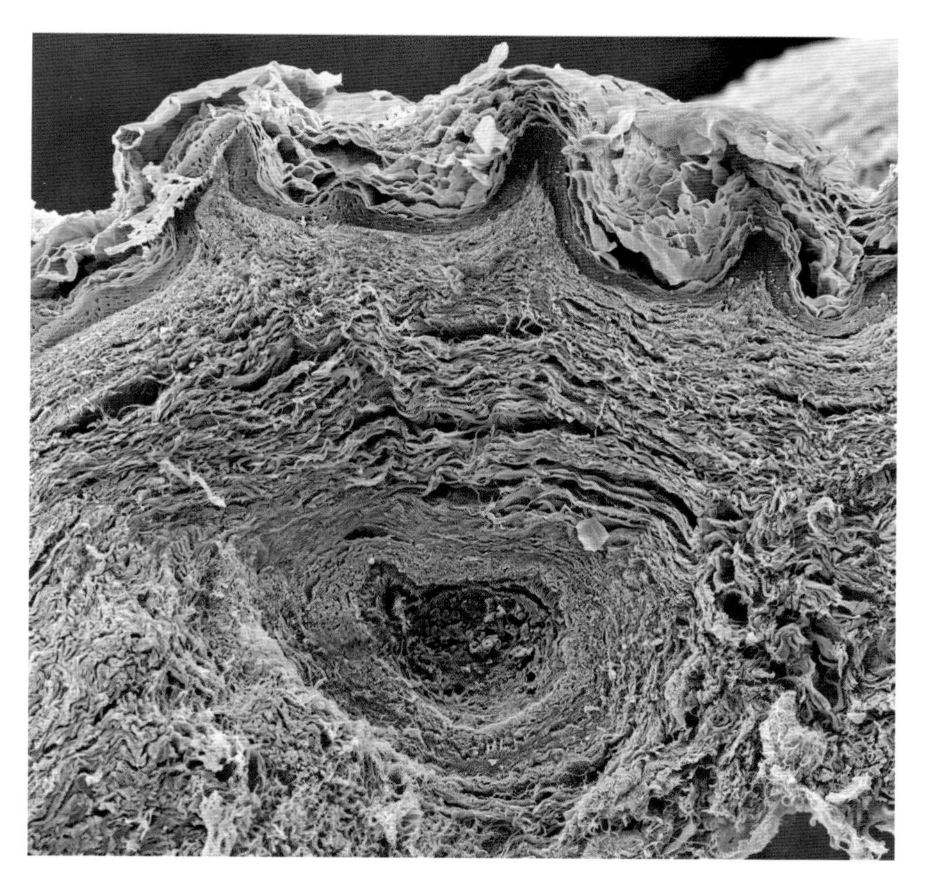

Keeping cool: a sweat gland (blue) lies within the dermis (grey-brown), deep below the rippled epidermis (red, with stratum corneum in yellow).

hair follicles. Some sweat glands are modified to make other substances, such as the wax glands in the ear. More strangely, the milk-producing mammary glands in the breasts are also altered sweat glands.

Hair

Hair follicles, cup-like structures forming sockets for hairs, are also found in the dermis. As well as being sockets, they are the place from which the hair grows. A group of cells at the bottom of the follicle divides and divides, pushing the lengthening hair up. Small hairs all over the body have little muscles attached to them. These are smooth muscle and part of the automatic systems of the body. When you are cold, signals from the

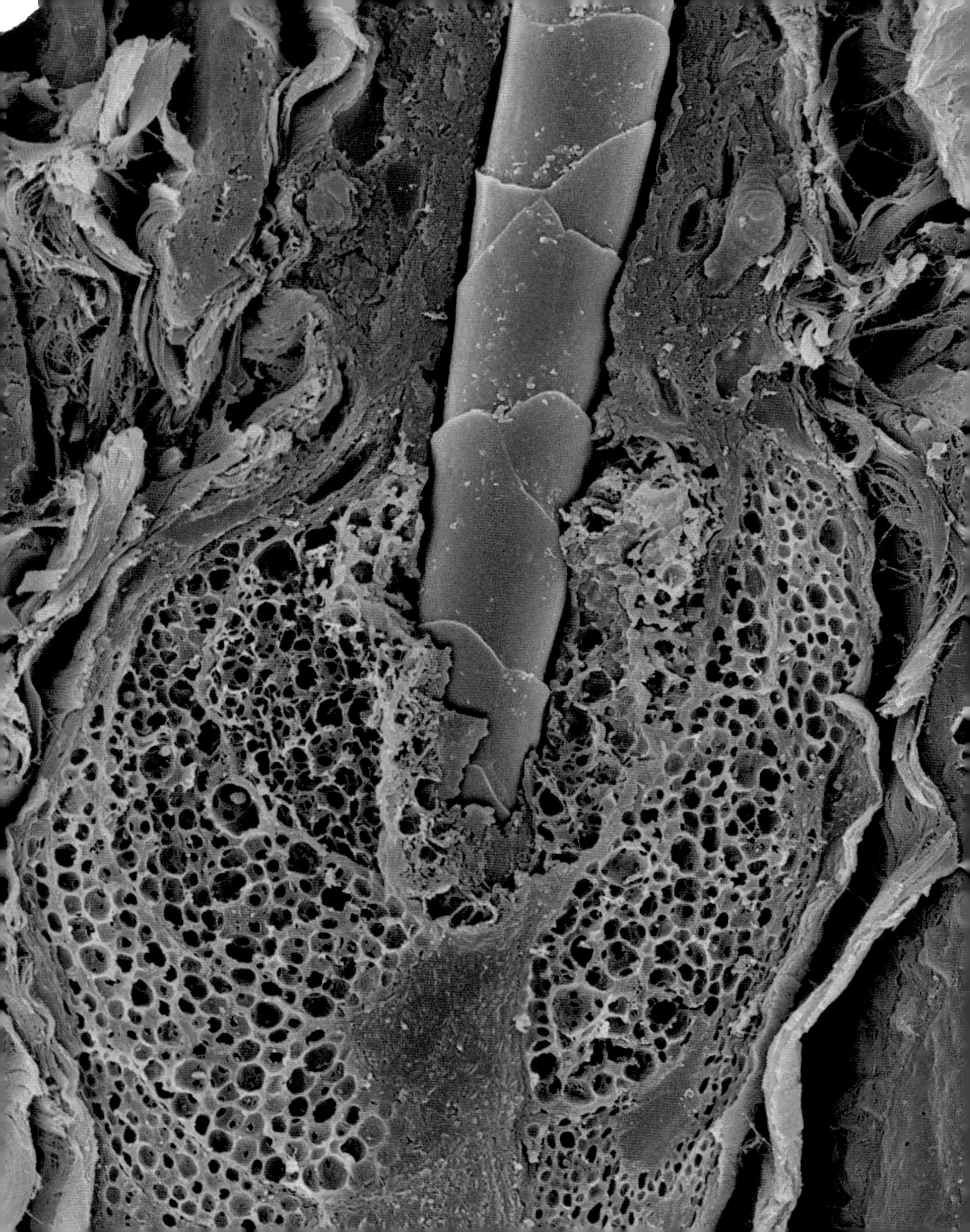

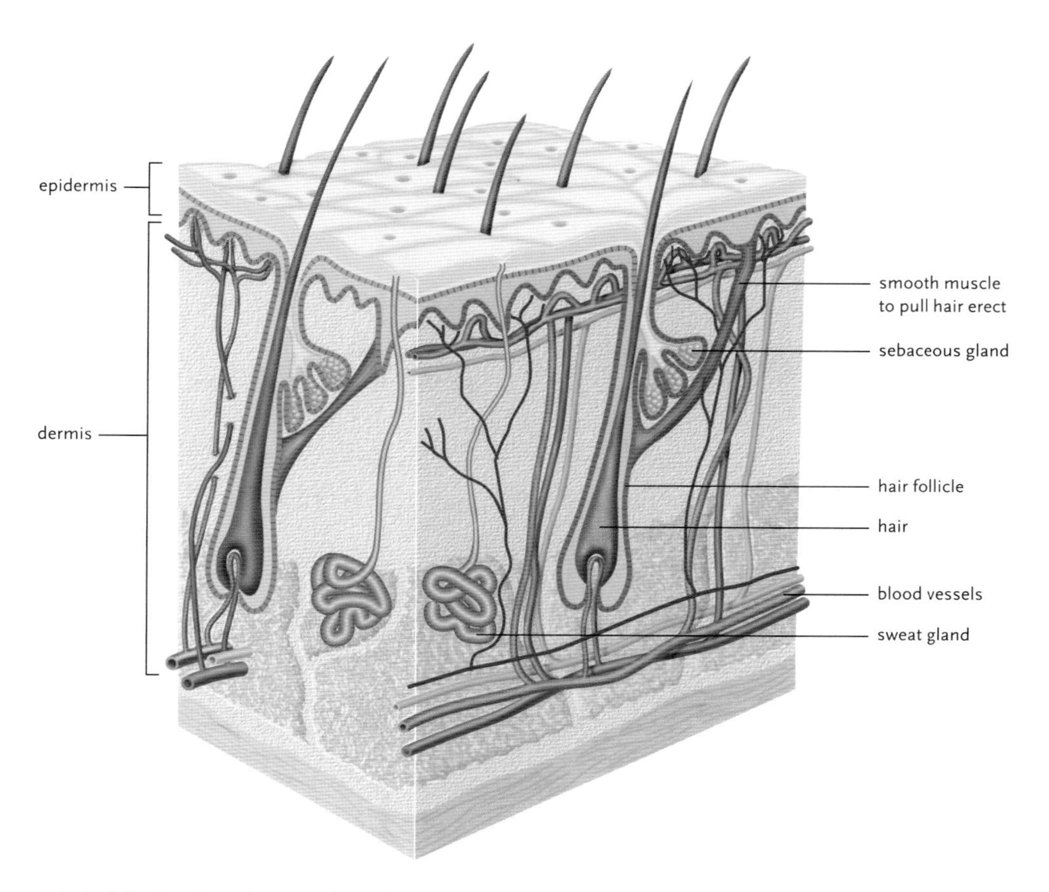

A block of skin: blood vessels, nerves, hair follicles, sweat and sebaceous glands packed into the dermis.

autonomic nervous system make these small muscles pull on the hairs, raising them. This system is the one that controls your response to fright; hence the sensation of 'the hairs on the back of your neck standing on end'. Making the hair stand on end has very little effect on keeping a human body warm, as our body hairs are so small, but it's a reminder of our evolutionary past. We were not always naked apes; our ancestors were hairy all over. Actually, we are still covered in hair everywhere (apart from our palms, soles, nipples, the end of the penis and the labia minora) but it is very fine. The body hair of women and children is much finer than that of men.

A hair (greenish yellow) emerges from its follicle, surrounded by the sponge-like tissue of a sebaceous gland (pink). The purple strands are part of the muscle that pulls the hair erect.

Apart from the fine hairs covering most of your body, there are areas where much coarser hair sprouts: on your scalp, brows, armpits and pubic areas – as well as on the face, chest and limbs in men. There are about 100,000 hairs on your head (if you have a full head of hair). Each one grows for about four years before falling out and being replaced, and nearly 100 hairs are shed each day. After the age of 40, the hair follicles generally start to wither and the hair begins to thin.

Sebum

The skin produces its own lubricant: sebum (Latin for 'grease'). This prevents the skin from drying out, keeps it supple and has an anti-bacterial effect. Sebaceous glands (the ones producing sebum) mostly open into the hair follicles and from there the sebum spreads out on the surface to soften, lubricate and waterproof the hair and skin. Sebaceous glands are stimulated by male hormones and become very active during puberty, causing the dreaded zits. A blocked sebaceous gland fills up with sebum, producing a 'whitehead', which oxidizes to become a 'blackhead'. If bacteria manage to get in and infect the trapped sebum, this produces acne.

Vitamin D

Although we are not plants and can't make sugar from sunlight, we do need sunlight to manufacture a very important product: vitamin D, which is essential for maintaining healthy bones. As well as interfering with calcium metabolism and leading to softened bones (rickets and osteomalacia), a deficiency in vitamin D can contribute to high blood pressure and insulin resistance. Getting some sunlight on to your skin is essential; a major cause of vitamin D deficiency is a 'sun-deprived' lifestyle.

The colour of disease

The colour of our skin depends primarily on its natural pigmentation and also the amount of blood flowing through the dermis, but skin colour can also indicate disease. A very pale skin may indicate anaemia or low blood pressure, while very red skin may be due to high blood pressure or

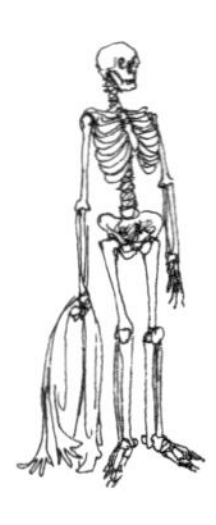

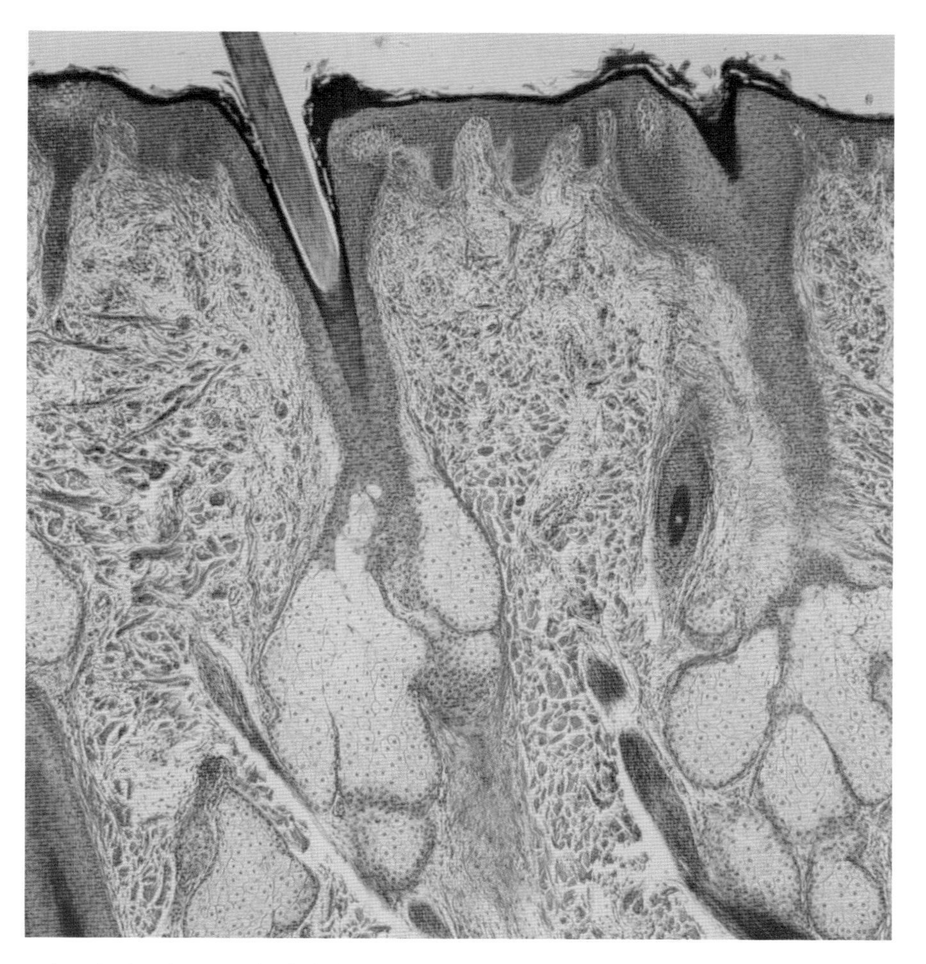

Scalp under the microscope: the shaft of a hair emerges (upper left) from a follicle. Sebaceous glands can be clearly seen as pale pink pockets in the lower half of this image.

inflammation. Well-oxygenated blood will turn skin pink, but poorly oxygenated blood may turn skin purplish; blue lips may indicate heart or lung problems, where insufficient oxygen is getting into the blood. Blood that escapes from blood vessels stains the skin: bruising. Other pigments may turn the skin a strange colour: excess bile pigment in liver disease produces jaundice and a more orangey-yellow colour can be produced by carotene. When I was a junior doctor, I remember seeing an entire orange family, because mum had decided that she and her children were going on an exclusively carrot-based diet.

Common skin problems, and how to prevent them

Healthy skin protects us from infection, mechanical and chemical damage and radiation. If the skin is damaged, this protection is impaired and further damage will ensue. Dry skin is an ineffective barrier, so it's important to keep your skin clean while preventing it from becoming dry. Skin can become inflamed either as an atopic response or through contact with irritants. Ultraviolet radiation can cause sunburn and skin cancer, as well as cumulative ageing effects.

Now we turn to the environmental factors that threaten the health of your skin and what you can do to combat them, and look after your skin.

Dry skin

Dry skin can be caused by diabetes or through inherited skin disorders like ichthyosis and atopic dermatitis. It can also be the result of exposure to chemicals, solvents and detergents. If skin dries out, the normal shedding of epidermal cells stops, because the enzymes that usually cut the cells loose can't work in the dry, and the old cells pile up on the skin's surface. Dry skin feels rough and flaky and can be uncomfortable, painful, itchy and stinging. Dry skin stops being an effective protective barrier: your skin becomes permeable and more susceptible to irritation and dermatitis or eczema.

Normally, the skin's moisture is kept sealed in by the barrier of the stratum corneum. This is much more than just an inert layer of dead packets of keratin waiting to be sloughed off. The overlapping, flattened cells contain a natural moisturizer and there are oily layers between the cells that help to seal in the moisture and prevent this natural moisturizer from leaching out. The stratum corneum is like a wall, where the cells are the bricks and the oils are the mortar.

If you want to clean up something oily, you'd probably use a detergent. Detergents, or surfactants, work by lowering surface tension and dividing oil droplets into smaller droplets – just as washing-up liquid breaks up grease and allows you to wash the smaller blobs of oil off your plates. Surfactant is

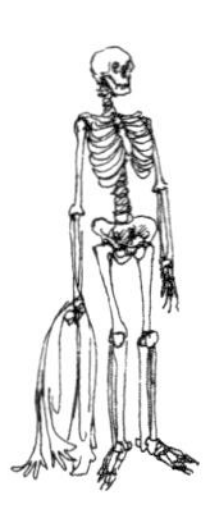

the main lathering ingredient of soap, which means that soap, very efficiently, strips the stratum corneum of its protective oils, allowing your natural moisturizer to be washed away. This stops the stratum corneum doing its job as an effective barrier: well-hydrated skin is a good barrier; dry skin is more permeable. The stripping of your natural oils can lead to irritation and the epidermis becomes inflamed and thickened, further reducing its effectiveness as a barrier. A vicious circle starts – this is the basis of skin conditions like dermatitis.

Products to use on your skin and hair: Should we all stop washing immediately? No, because soap is also very good at washing away harmful bacteria, so it's important for general skin hygiene. But it does mean that, if you tend to have dry, itchy skin, you might be using too harsh a soap or cleanser and it's probably worth looking for something a bit milder. Some soaps contain moisturizers or emollients, which can help to replace some of the natural moisturizers that the detergent will have stripped out. It's about striking a balance: if you use a milder soap, less moisture will be lost from your skin in the first place, so you shouldn't need to use as much moisturizer. The surfactants in most liquid body washes and shower gels (and in facial cleansers in particular), are generally less harsh, and therefore less drying, than those in soap bars.

A basic moisturizer is a mixture of oil and water that can help hydrate the skin and replace some of the oils between the cells of your stratum corneum. A better moisturizer also provides moisturizing substances – humectants – to your skin, to replace the natural moisturizers. There's some evidence that moisturizers with humectants are better than an oil-and-water mix in preserving the barrier function, as the latter may make the skin more susceptible to irritation.

Your skin changes over the year: the stratum corneum tends to be drier in the cold winter and naturally more moist in the warm, humid summer. So, you may find that you get dry skin in winter and need to apply moisturizer daily, but that you don't need it so often in the summer. Our skin also responds to our normal biological cycles, particularly to women's monthly changes in progesterone and oestrogen levels. Your skin tends to be driest during your period and oiliest in mid-cycle, just after ovulation.

The skin also changes as we get older; there are inevitable changes in the structure of our skin as we age. The dermis becomes old and tired, and the epidermis gets thinner and sheds cells more slowly. Skin gets drier as we get older: the stratum corneum isn't as good at holding on to water and the sebaceous glands are less productive than in young skin. In the dermis, blood flow is reduced and fragmentation of collagen and fewer elastin fibres result in wrinkles. Our immune system stops working as efficiently as we get older and this, combined with the impaired barrier function of the stratum corneum, means that skin infections become more common. Keeping the skin clean and preventing the stratum corneum from drying out is very important in reducing your risk of getting skin infections. It doesn't matter how old you are: the stratum corneum will always work best and look best if it is not dried out.

The best way to avoid dry skin is to use mild soap or cleanser and a moisturizer when necessary. You might as well stick with what you're using already if you've experienced no adverse effects. If your skin is overly dry, try using a milder cleanser to preserve more of the skin's natural oils and moisturizers, before bumping up the amount of artificial moisturizer you're sticking on it. Avoid using powders, like talcum powder, as these can cause the skin to dry out by absorbing the protective oils from the stratum corneum. This sounds very simple – and for most people it really is that simple most of the time. But if your skin is very dry and these simple measures don't fix it, it's certainly worth going to see your doctor about it: dry skin can make infections more likely and there may be other underlying medical problems causing it.

Your hair needs the same sort of care as your skin. It's a good idea to keep it clean, unless you don't mind it becoming a pleasant home for all manner of pungent bacteria. In choosing haircare products, you have to strike the same balance as you do with your skin: using a strong detergent shampoo will divest your hair of all its natural oils, and you'll find that you have to replace these with a moisturizing conditioner. If you use a milder shampoo, you may

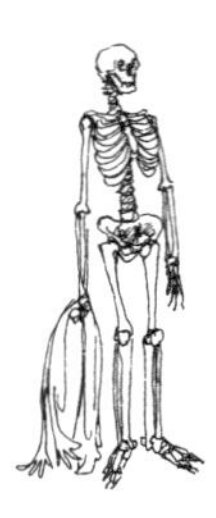

Split ends: hair damaged by bleach, seen under an electron microscope.

find that you don't need any conditioner. It's important to remember that hair is dead, so any amount of vitamins you throw at it won't do an iota of good. Vitamins (Latin for 'life') play an important role in the processes that go on in living cells. But the elixir of life is not much use to dead cells.

Dermatitis and eczema

These terms both refer to inflammation of the skin, although 'eczema' is often used as shorthand for atopic eczema. Inflamed skin doesn't work well as a barrier and dermatitis is often made worse by bacterial infection.

Atopic eczema, like asthma, might develop due to a lack of exposure to infections in infancy. Episodes of atopic eczema can be brought on by allergens such as furry animals or food allergens and by direct contact with strong detergents and other chemicals. Stress and anxiety can also exacerbate eczema. Avoiding irritants, especially soaps, cats and dogs, wearing cotton clothes and not getting too hot can all help to prevent exacerbations of eczema. For children, the good news is that most will grow out of their eczema by the time they are teenagers.

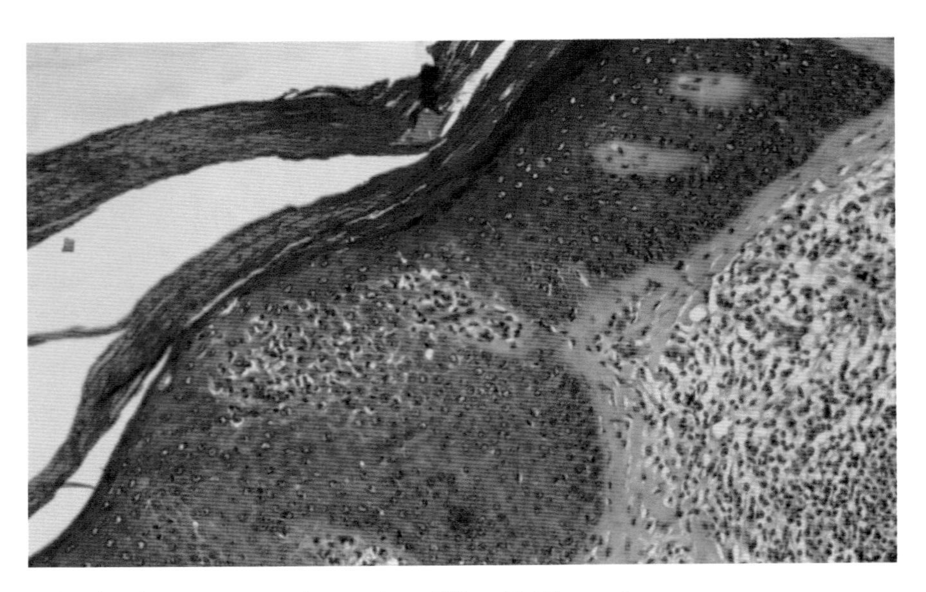

Peeling skin: the stratum corneum is separating and lifting off in this case of eczema.

'Winter eczema' is more common in the elderly and is probably due to the natural tendency of skin to be drier in winter and old age, perhaps aggravated by use of strong soaps. 'Contact eczema' is caused by an irritant – often detergents, soaps or bleach – coming into contact with the skin. Sometimes repeated exposure can lead to an allergic reaction set off by contact with nickel (in jewellery), cement, latex, perfumes and plants.

It's important to be careful about what you bring into contact with your skin. Wear gloves when you're handling detergents, bleach or cement. It's also important to be aware that some of the things we put on our skin to look after it might actually produce irritation or an allergic reaction. Skin cleansers and moisturizers might be an important part of looking after your skin and keeping it healthy, but they can cause problems. Introducing a concoction of chemicals on to your skin might cause irritation or eczema. It isn't just synthetic chemicals you need to watch: there's plenty of evidence that increased use of 'natural' or botanical extracts has caused problems either directly or by their interaction with other compounds.

Adverse reactions to personal care products (skin- and hair-care products, deodorants, cosmetics, perfumes and nail polish) are very

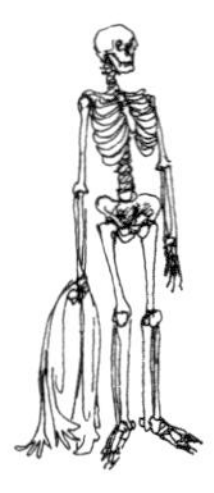

common: nearly one in four women and about one in seven men are affected each year. These adverse reactions vary from mild irritation of the skin to quite nasty reactions, with reddening of the skin, burning or itching sensations, blistering and bleeding.

Just about anything you put on your skin has the potential to cause a reaction, especially with repeated applications. Having said that, the core ingredients of moisturizers, the humectants and oils, very rarely cause irritation or contact allergy (although humectants like urea may break down keratin and weaken the skin's barrier function). The most common allergens are perfumes and preservatives. If you've got sensitive skin and you've had adverse reactions to a product in the past, it's probably a good idea to avoid perfumed products and go for simple formulations. Various dyes used in cosmetics can also produce adverse reactions, so it's a good idea to avoid coloured products if you've got sensitive skin. Oil-based substances tend to permeate the skin quite easily, even with a well-functioning barrier. This means that things like essential oils can literally 'get under your skin'. It's not clear what sort of effects tiny amounts of these chemicals might have on your body, but the simple fact they can get into the skin means they have the potential to cause irritation and sensitization, and more and more adverse reactions to essential oils are being reported.

Photo-ageing, sunburn and cancer

Managing your exposure to sunlight is very important. The skin needs sunlight to make vitamin D, but we need to be mindful of the dangers of too much sun: an immediate risk of serious sunburn and the longer term risk of skin cancer. Exposure to sunlight also has an ageing effect on the skin.

Photo-ageing: Ultraviolet radiation has a cumulative, ageing effect on skin, sometimes called 'photo-ageing'. It causes damage to collagen and elastin, the proteins in the dermis that give your skin its structural integrity, toughness and elasticity. Signs of photo-ageing include wrinkles and increased stretchiness, changes in the pigmentation of the skin, telangiectasia (small, spidery clusters of dilated capillaries), leatheriness

HEALTHY SCEPTICISM: MOISTURIZERS THAT REVERSE THE AGEING PROCESS

What about those incredibly expensive products that claim to reverse the ageing process, unwrinkle wrinkled skin and make us all look (or stay looking) more youthful? In the plethora of adverts we read in magazines or see on television, dermatologists agree that skin-care companies are inclined to exaggerate, and even make outrageous claims about the effectiveness of their products. Cosmetics companies lace their ads with impressive-sounding pseudo-scientific terminology. Most scientists and doctors have come round to the idea that they should be trying to explain things simply (without dumbing-down), so it's quite ironic that others are trying to use dense scientific-sounding terminology to impress people. It has to be said, there's also a lot of unscientific posturing on the other side. Loads of articles and books play up the potentially harmful effects of lotions and cosmetics; while I applaud their general message that the cosmetic companies may not be as honest as you thought they were (depending on how honest you thought they were in the first place), some have gone a bit too far the other way.

Using moisturizers on dry skin will certainly help keep your skin looking healthy, (if not younger), as well as looking after its barrier function. The barrier function means that most things you put on your skin don't penetrate very deeply (and you wouldn't want them to). Remember that the epidermis is constantly shedding cells and being

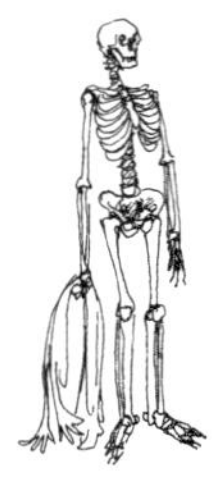

renewed; so while mild cleansers and moisturizers may help to keep the stratum corneum healthy, any effect they do have is temporary.

Some creams that claim to have specific anti-ageing effects contain exfoliants (a word which comes from the Latin 'to strip off leaves'). These are acids, for example, alpha-hydroxy acids (AHAs), that etch away the outermost layers of your skin. About 30 years ago, AHAs were found to be very good at stripping away the outer layers of skin in patients with ichthyosis, who suffer with excessively dry, scaly skin. They started to appear in moisturizing products, but just because they are useful in treating clinically dry skin doesn't mean they are going to be beneficial if you use them every day on normal skin. These skin-peeling products might actually cause the skin to thicken, although this may not be noticeable until you stop using the exfoliant. While they may temporarily make your skin look smoother, they also remove part of your skin's natural barrier, so that anything else you put on your skin can penetrate more deeply and cause irritation and ultraviolet rays may be able to reach further into the skin. Other anti-ageing creams contain retinoids, like tretinoin, which appear to work by accelerating collagen breakdown. This seems a rather destructive way of achieving younger-looking skin and these substances can also sting and irritate the skin. There are anti-ageing creams that work in a less aggressive way, and some studies suggest that products containing antioxidants, like vitamin C or co-enzyme Q, may be able to slow down photo-ageing.

and an increased risk of skin cancer. Photo-ageing affects lighter-skinned people more than darker-skinned people. This is perfectly understandable; light skin has less melanin, our natural sunscreen, than dark skin. So the dermis of people with darker skin has a natural advantage in photo-ageing. Artificial sunscreens, although very good at preventing sunburn, haven't been tested for long enough to gauge their effectiveness in preventing photo-ageing.

Sunburn: Sunburn is well named: the skin reacts to ultraviolet radiation in the same way as it does to intense heat. The immediate and obvious effect is a reddening of the skin – erythema – as the capillaries in the dermis widen to bring more blood to the surface. At a cellular level, cells in the middle of the stratum corneum are shocked into halting production of the natural moisturizing compounds. This creates a dried-out 'plane of weakness' that gradually moves up through the stratum corneum and eventually causes the whole layer above it to slough off. Even at lower levels of exposure, skin that is routinely exposed to sunlight becomes dry.

Skin cancer: Exposure to ultraviolet radiation doesn't just stop cells producing proteins, it can cause damage to the DNA of your genes. Mutations in the genes can ultimately cause skin cancer, including its most serious form, malignant melanoma. Sunscreen has been at the forefront of public health campaigns about limiting exposure to ultraviolet radiation and reducing the risk of skin cancer. But you shouldn't depend on it. Recently, concerns have emerged that people may be over-relying on the protective value of sunscreen and have been staying in the sun for longer than they otherwise would. This might explain why some studies have found that sunscreen use is actually associated with a higher risk of skin cancer. Sunscreen doesn't have talisman-like properties; it can't make you totally invincible and impervious to sunburn and skin cancer. It's best to start by limiting your exposure to sunlight and to use protective clothing, sunhats and sunscreen when you are in the sun. Remember that you must re-apply sunscreen, especially if you're on the beach and you've been swimming or you have towelled yourself off.

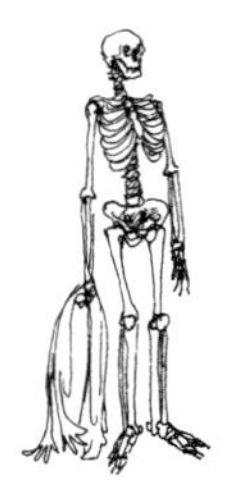

Malignant melanoma: cancerous melanocytes (orange) run wild, multiplying rapidly and invading the epidermis (green); magnified 30,00 times.

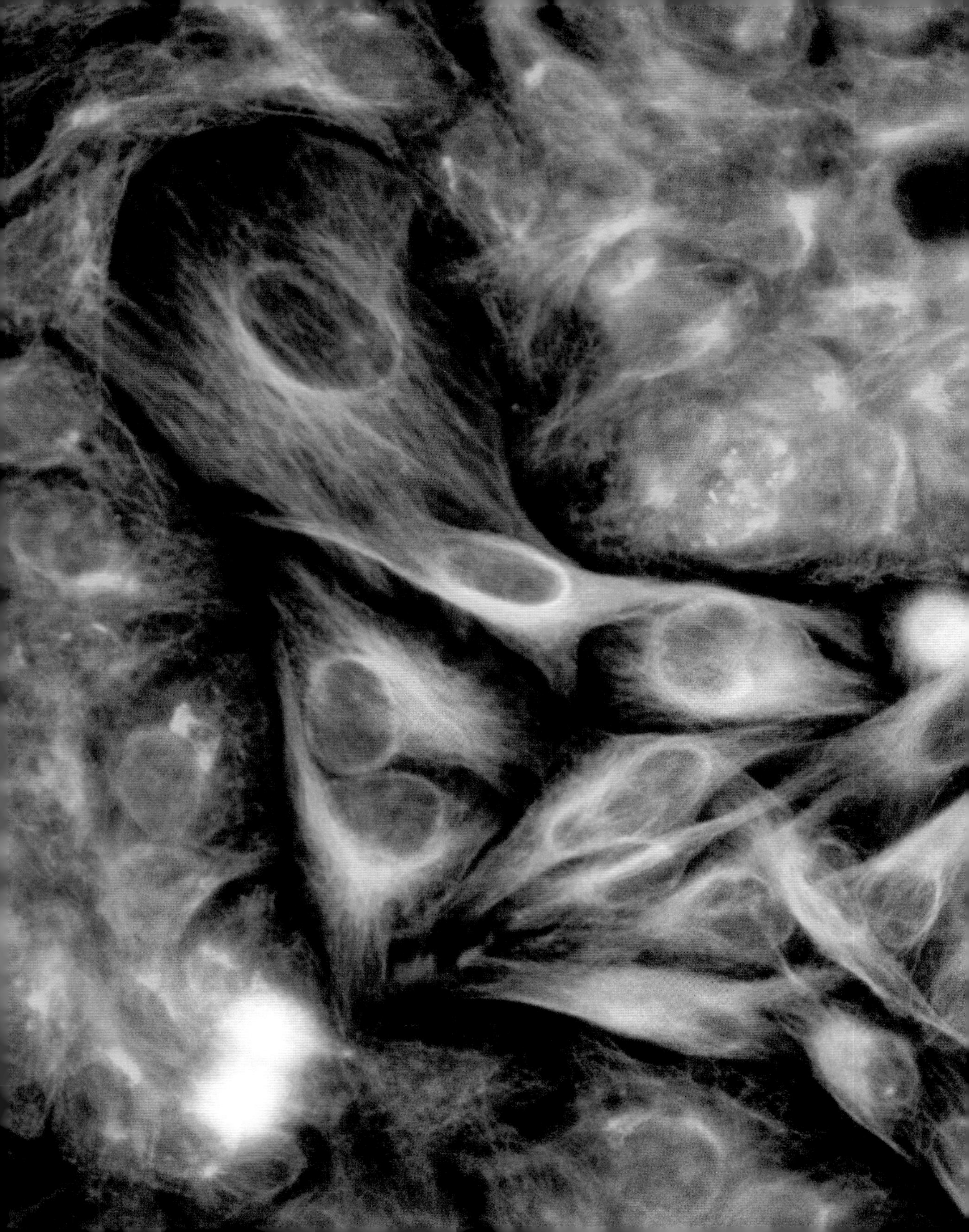

Complete sun avoidance might seem like the best plan – but remember vitamin D. You need to balance your exposure to sunlight with making sure that you get enough sun to make vitamin D. Cover yourself up entirely from head to toe or stay inside all day and your skin will struggle to make enough. And it's very hard to make up the difference with diet or supplements.

Being sensible about exposure to ultraviolet radiation shouldn't stop you going outdoors. Avoid too much exposure to sunlight: seek out the shade, especially when it's very hot (between the hours of 10 a.m. and 4 p.m. in the summer), wear long-sleeved tops and use sunscreen with an SPF of 15 or more if you've got to be in the sun. This will reduce your risk of skin cancer and help you avoid the most damaging effects of photo-ageing. It's tricky in a country like the UK, where occasional bursts of sunshine are so welcome. People who have become pale under clouded skies rush out to sunbathe, but this is the worst thing to do. It's like binge-drinking for your skin: pale skin isn't ready for harsh sun and quickly burns. Intense and intermittent exposure to ultraviolet radiation is the most dangerous; the most likely to cause skin cancer. Sunburn is linked to skin cancer. If you've got pale skin that burns easily, you really need to be especially careful. This is very important for children: just don't let them burn.

As well as protecting yourself from the outside, you can protect yourself from the inside: there are naturally occurring substances in the diet that can reduce your risk of developing skin cancer. Eat plenty of fruit and vegetables to stock up on the antioxidants which mop up the free radicals that can cause damage to DNA.

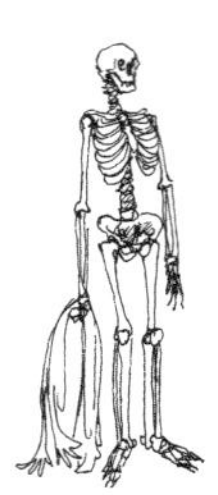

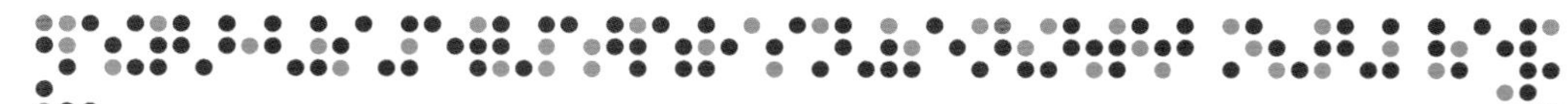

Five ways to keep your skin healthy:

- Limit your exposure to strong sun and ultraviolet rays to reduce your risk of sunburn and skin cancer.
- Keep skin clean to avoid infections; don't share towels or flannels with someone who has an infection.
- Avoid dry skin: use mild cleansers and a moisturizer when necessary, to preserve the barrier function of your skin.
- Avoid known allergens and irritants; choose skin products with simple formulations and minimum perfumes and preservatives.
- Eat a healthy diet with plenty of antioxidant-rich fruit and vegetables – and don't smoke!

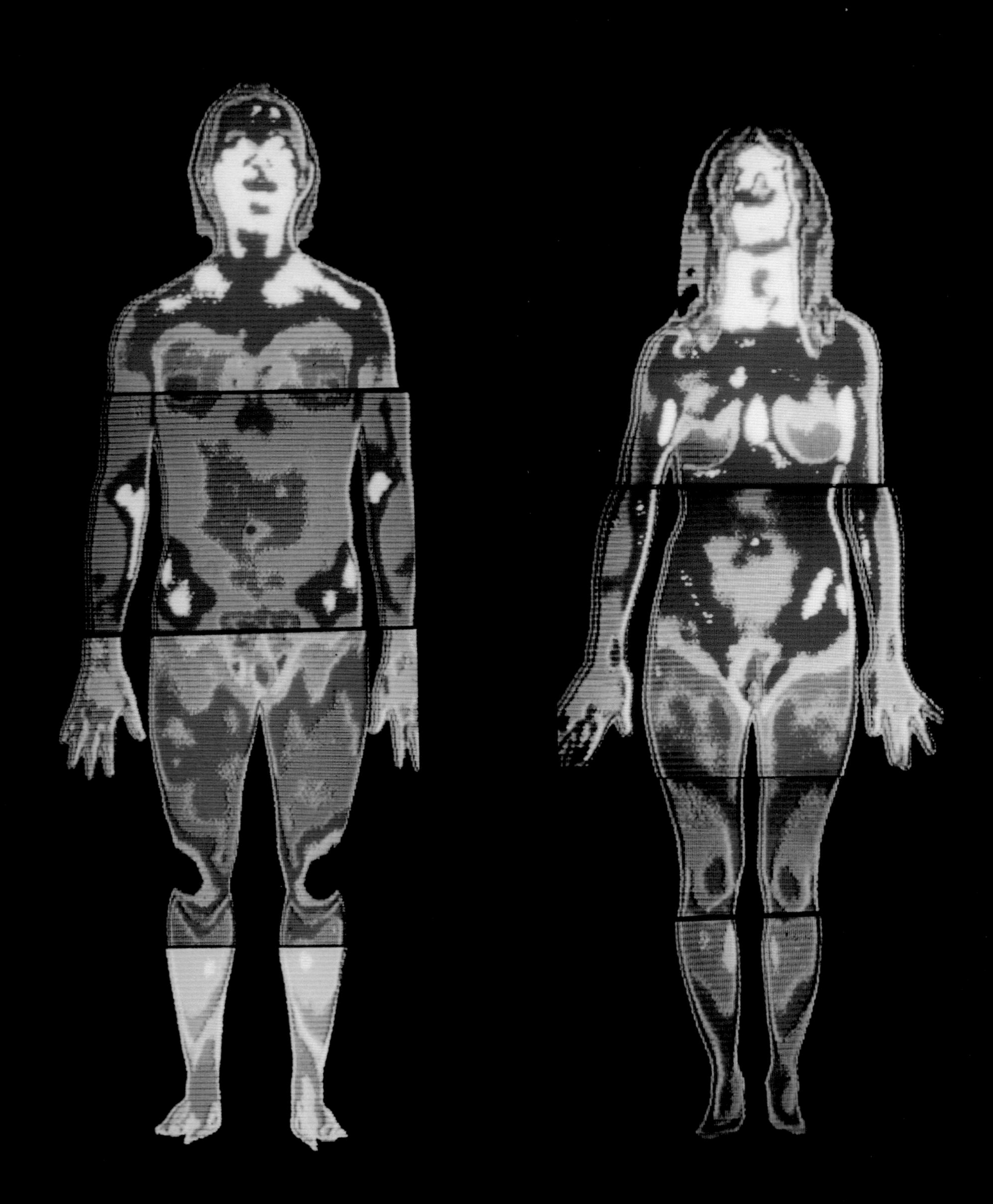

ELEVEN

THE WHOLE BODY

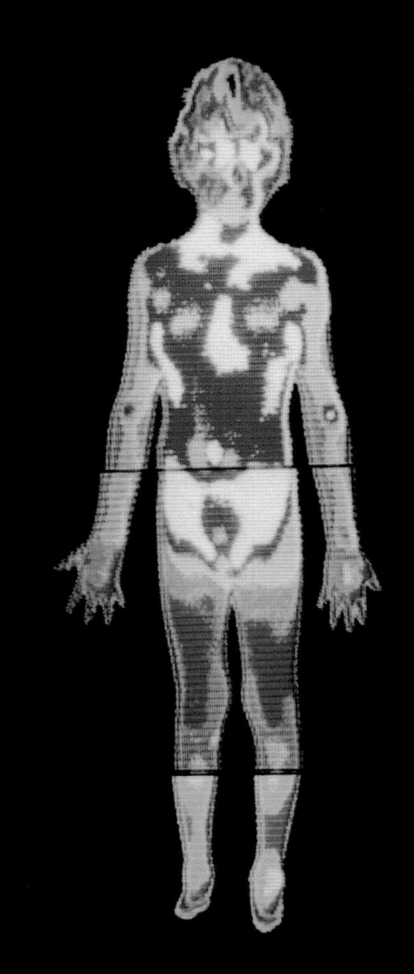

B

Each of us has our own, unique model of the human machine, with its personal idiosyncracies and imperfections. And none of us leads the 'perfect lifestyle', but the human body, like any other machine, will carry on working better for longer if we try to look after it. We all forget to oil it occasionally, top it up with water, or to cover it over to stop it rusting. We tend to take it for granted, but with a bit of thought, we can keep it in good working condition.

Some of the things that go wrong with it are inbuilt design faults, others are accidental. But there are lots of problems that we can take steps to avoid, which is what the health tips in this book have been about. At the end of the day, it's best to treat your body as a whole, and adopt a diet and lifestyle that will keep all your organs working at their best. They will all appreciate the benefits of a healthy diet and exercise.

Previous pages **Thermograms of a man, woman and an eight-year-old boy. Temperature is colour coded, from red (hot), through yellow, green and blue to purple (cold).**

Left **Vitruvian Man: Leonardo da Vinci's famous study of the ideal proportions of a man; but in reality, we are all variations on a theme; the human body is never 'perfect'.**

A 'final common pathway': free radicals versus antioxidants

In this book, I have concentrated on the ways in which you can look after the health of your organs, and your entire body, by lessening the impact of environmental factors, whether these come as part of your diet, in your level of physical activity or in the chemicals and radiation that you come into contact with. Many of the damaging factors that I have identified have a final common pathway, a shared mechanism by which they wreak their destructive effects on the body. Paradoxically perhaps, this pernicious common pathway involves a substance that we think of as life-giving, stimulating and energizing: oxygen. That oxygen should be an agent of decay and destruction in the human machine is perhaps not so surprising; it is, after all, what causes other machines to rust.

Humans, like every other oxygen-using organism on the planet (and that's the vast majority of them), have a love–hate relationship with oxygen. It keeps us alive and allows us to free the energy from fuels we take in, but it's also capable of killing us. When food materials are oxidized in the mitochondria, electrons are added to oxygen molecules, one by one. The end product, when hydrogen ions have also been added, is water. But on the way from the stable oxygen molecule (O_2) to two molecules of water (H_2O), free radicals are produced.

These free radicals are unstable, reactive species of oxygen: 'superoxides', where an unpaired electron has been added. The unpaired electron is the source of the free radical's destructive power. It's in desperate need of a partner; it's like an embittered mistress in search of a marriage to wreck. It will steal that electron from any molecule it can, with disastrous effects: disrupting proteins, lipids, DNA – and the enzymes that should normally repair DNA. Free radicals will even disrupt the very enzymes that are there to neutralize free radicals: antioxidants. Free radicals can spread their destruction by creating even more dangerous molecules, such as hydrogen peroxide, causing a cascade of chaos in the cell and resulting in the death of the cell. Sometimes the body harnesses the destructive power of superoxide

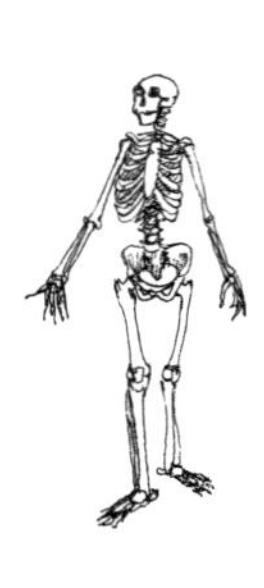

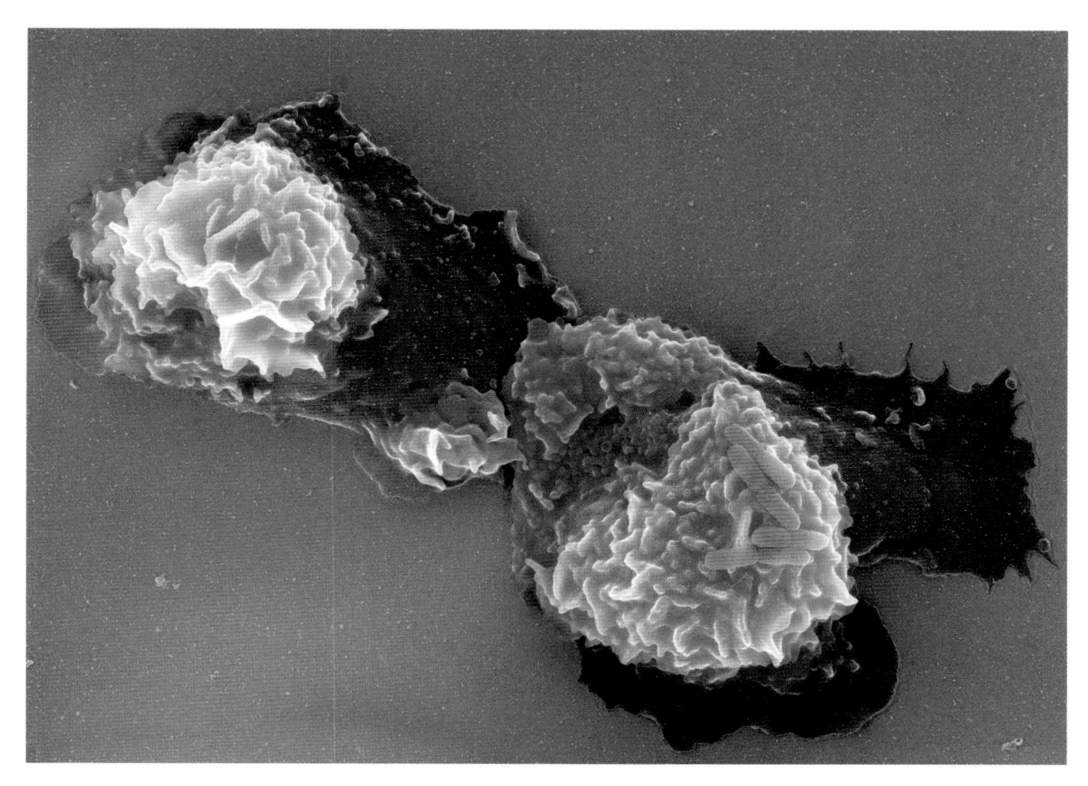

Imminent demise: these dysentery-causing *Shigella* bacteria (the orange rods) are lying on the surface of a neutrophil (a type of white blood cell), which is about to engulf them and destroy them with free radicals.

free radicals to its own benefit: white blood cells – neutrophils – deliberately create superoxide to kill invading bacteria. But handling such a deadly weapon is always dangerous: the free radicals generated in the immune response may harm the body's own cells as well as the bacteria.

Damage to cells and tissues can disrupt the neatly compartmentalized process of oxidation in the mitochondria, so that oxygen merely halfway through the process of having electrons added is released – as more free radicals. When tissues are deprived of oxygen, strange enzymes form, which make free radicals when the oxygen supply is restored. The direct damage caused by ischaemia in a heart attack is made even worse by the release of free radicals afterwards. Environmental factors, such as ionizing radiation (including ultraviolet radiation), polluting chemicals and cigarette smoke can also cause the production of free radicals.

Can we avoid free radicals?

It's not possible to avoid free radicals entirely, as they're part of the normal process of oxidation in our cells, and used in our immune response. What we can do is to limit our exposure to pollutants and radiation, which can create more free radicals, and make sure that our bodies have enough antioxidants to keep the free radicals under control. Antioxidants are a diverse lot: some stop free radicals being made, some mop them up once they're produced and some repair the damage that they cause. Some are made in our bodies, but others, such as vitamins C and E, we get from our diet.

Vitamins C and E are scavenging, mopping-up type antioxidants – they find free radicals and neutralize them. Our diet also supplies the fuels that get oxidized in our cells to release energy, and not all fuels are equal in the free radical stakes. Polyunsaturated fatty acids from meat are more susceptible to attack by free radicals than the monounsaturated fatty acids that predominate in vegetable oils. Olive oil is a double enemy of free radicals: it is composed mainly of monounsaturated fatty acids and contains antioxidant flavonoids and phenolic compounds.

Oxidative stress

'Oxidative stress' is a term used to describe what happens if the balance tips, when there are too many free radicals in a cell to be neutralized by the cell's antioxidants. The cell can try to make more antioxidants but, eventually, oxidative damage will occur. Some diseases might be caused directly by oxidative damage, while others might cause damage to cells and tissues that in turn leads to oxidative damage.

Oxidative stress may be the problem lying right at the core of coronary heart disease: fatty plaques of atheroma form in arteries, as white blood cells absorb cholesterol from the bloodstream. The cholesterol they absorb is oxidized 'bad' (LDL) cholesterol. Free radicals may be responsible for oxidizing the cholesterol, kicking off the sequence of atherosclerosis. Oxidative stress may also play a key role both in the cause of diabetes and in its secondary effects, such as the increased predisposition to atherosclerosis. It may also be implicated in Alzheimer's disease.

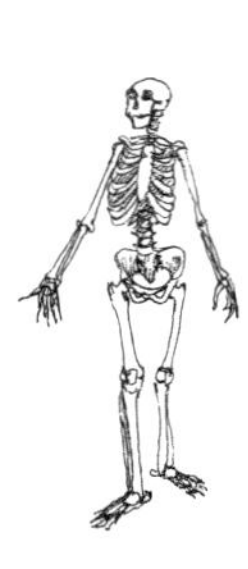

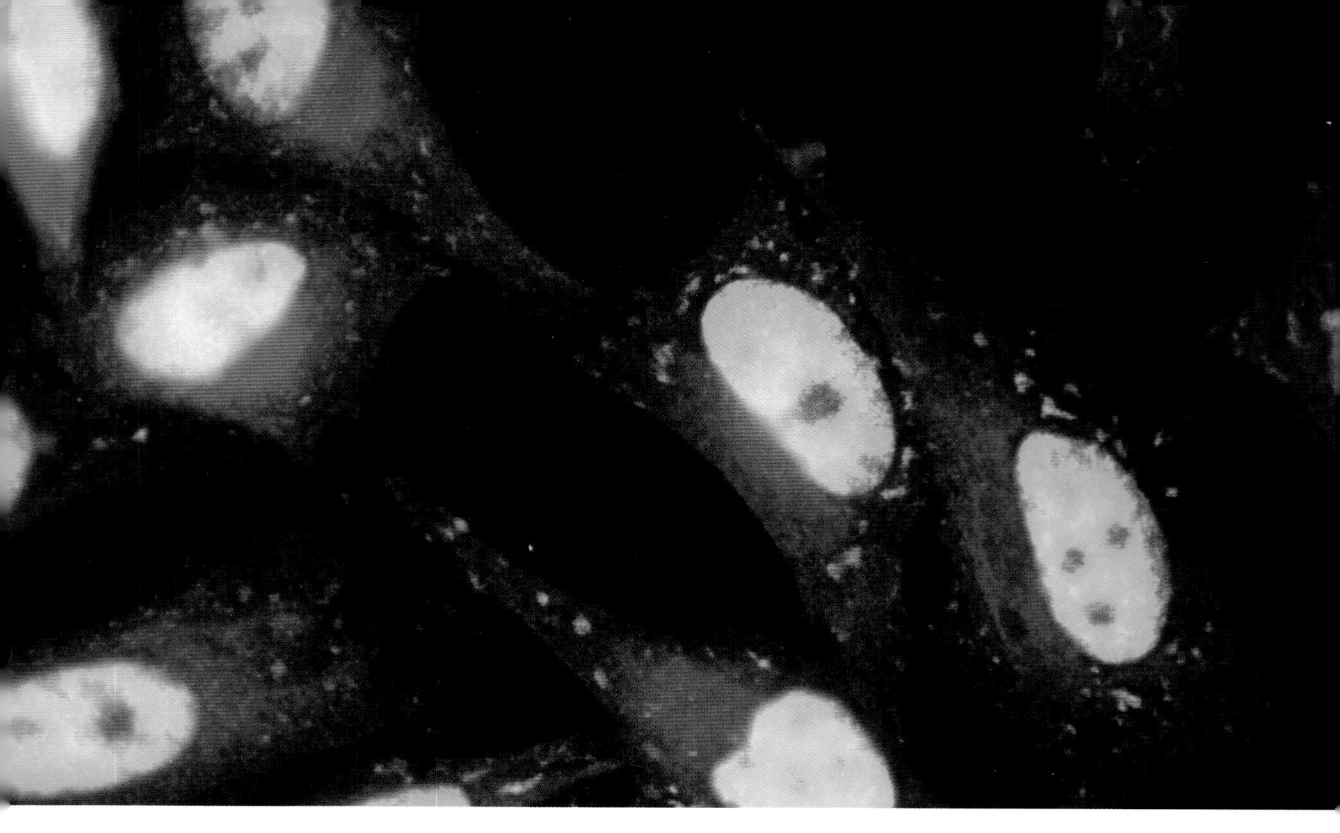

Prostate cancer cells: the green ovals are their nuclei, but the glowing orange granules are a sign that these cells are suffering oxidative stress.

There is evidence to suggest that the cells of the immune system work better if they have access to plenty of antioxidants: neutrophils seem to be protected from destruction by their own free radicals by antioxidants. Vitamins C and E are the guards to the neutrophils' sword: they stop the white blood cell from being cut by its own weapon. Antioxidants, by decreasing inflammation, also appear to have a positive effect on long-standing auto-immune diseases, such as rheumatoid arthritis and inflammatory bowel diseases. And they protect against cataracts and damage to the retina of the eye.

Like most other diseases, cancer involves interplay between innate, genetic factors (such as genes that predispose an individual to developing cancer) and environmental factors, some of which may cause mutations in their DNA. Early studies of diet and cancer focused on the potential carcinogens (cancer-causing compounds) that could be found in foods but, by the 1970s, the emphasis had shifted to identifying substances in food that

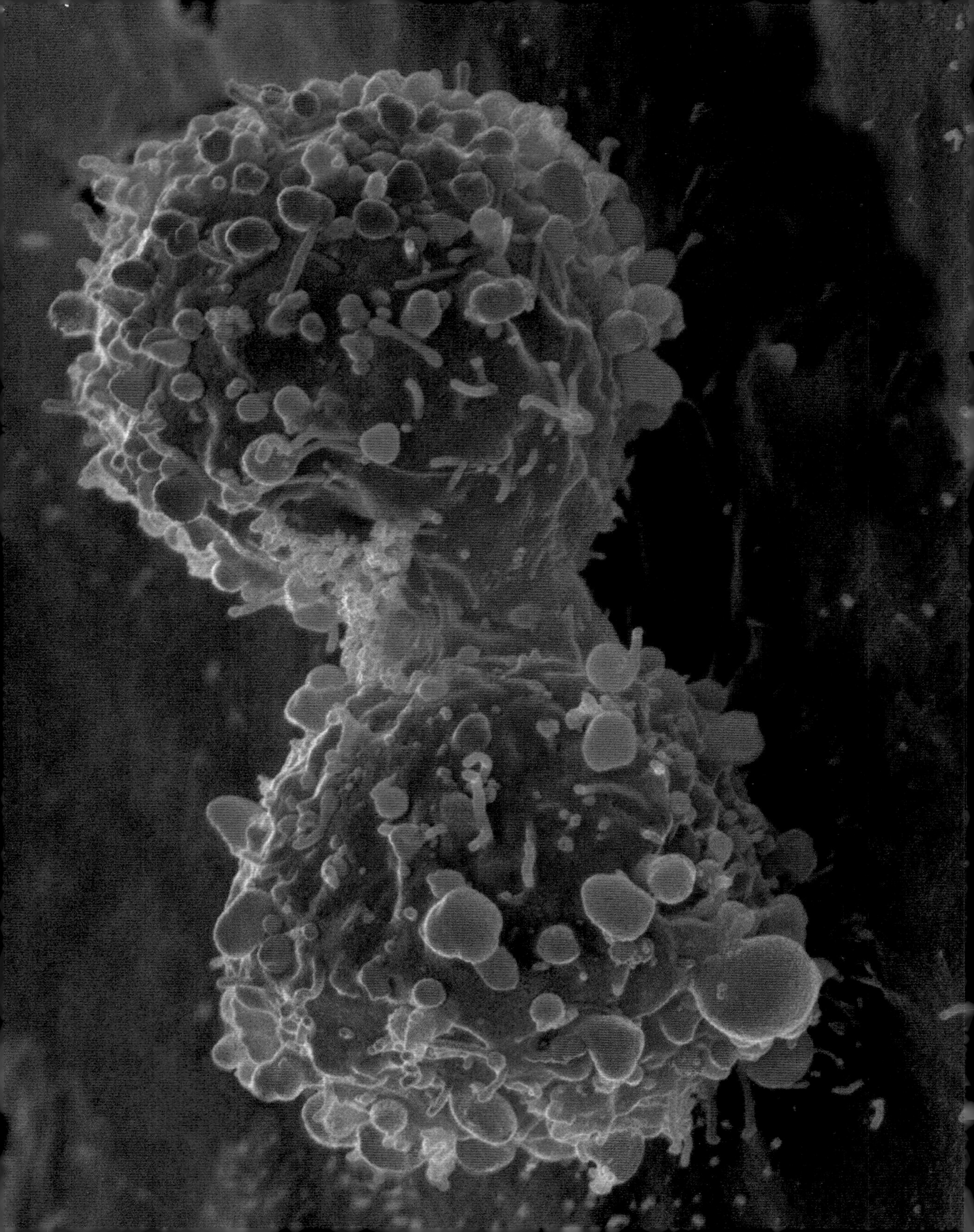

might protect against cancer. Free radicals can directly damage the DNA in cells, as well as damaging the enzymes that repair DNA. Damaged DNA may turn the cell into a tumour cell, which forgets its origin (as a specialized skin cell, lung cell or gut cell) and starts to multiply uncontrollably. It seems logical that antioxidants should be able to reduce the risk of cancer caused by free radicals. Epidemiological studies seem to support this: a diet rich in fruit and vegetables, and therefore full of antioxidant vitamins, flavonoids, phenols, and trace elements like selenium and zinc (which are components of antioxidant enzymes), reduces a person's overall cancer risk by between 30 and 50 per cent. People with a very low consumption of fruit and vegetables are twice as likely to get cancer as those eating plenty.

Considering the range of cancers that can affect just about every tissue and organ in the body, it's surprising that the dietary good guys and bad guys are the same for all of them. Vegetables and fruit seem to reduce the risk of cancer everywhere; smoking, alcohol and obesity to increase it.

In Britain, we eat less than half the amount of fruit and vegetables that people in Mediterranean countries do, and our rate of cancer is high. Too much red and processed meat is linked to higher rates of cancer (and obesity). Excessive consumption of alcohol is a risk factor in oesophageal, liver and breast cancer. As for where we get our antioxidants from, the evidence shows that it's diet that counts: supplements don't seem to have much beneficial effect and some may even be harmful in high doses.

It's often very difficult to pin down which specific nutrients are having the beneficial effects and whether it's down to a combination of antioxidants. Until we know more – and probably even when we do – the best advice has to be to get your antioxidants the 'natural' way by eating them in fruit and vegetables. You can be absolutely sure that, whatever specific nutrient or combination of nutrients it is that does the trick, you're going to be doing yourself a lot of good if you include plenty of fruit and vegetables in your diet.

Fruit and vegetables have actually been tested to see which have the highest levels of antioxidant effect. Fruit, nuts and beans come in at the top of the league: pecans, kidney beans, walnuts, hazelnuts, cranberries, prunes,

Unregulated expansion: a lung cancer cell just about to divide in two. A feature of cancer cells is rapid and uncontrolled cell division – cancer grows quickly.

Aspects of diet and lifestyle that affect the risk of developing cancer
(adapted from Cummings & Bingham, 1998)

Type of cancer	Increased risk	Decreased risk
Lung	smoking, meat, alcohol	physical activity, fruit and vegetables (but none of these can combat the increased risk from smoking)
Colorectal	obesity, red meat, alcohol	physical activity, fruit and vegetables, fibre
Stomach	salt, dried meat and fish	fruit and vegetables, vitamin A, C
Oesophagus	alcohol, smoking	fruit and vegetables
Breast	obesity, red meat, alcohol	physical activity, fruit and vegetables
Endometrial	obesity	
Cervix	HPV infection, smoking, obesity	fruit and vegetables, vitamin A, C, folate
Prostate	red meat, fat	vegetables, vitamin E
Bladder	smoking	fruit and vegetables
Pancreas	smoking, red meat	fruit and vegetables, vitamin C
Liver	alcohol	

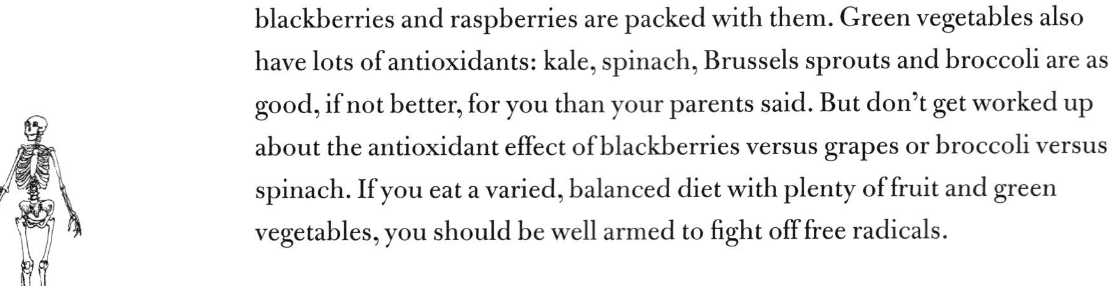

blackberries and raspberries are packed with them. Green vegetables also have lots of antioxidants: kale, spinach, Brussels sprouts and broccoli are as good, if not better, for you than your parents said. But don't get worked up about the antioxidant effect of blackberries versus grapes or broccoli versus spinach. If you eat a varied, balanced diet with plenty of fruit and green vegetables, you should be well armed to fight off free radicals.

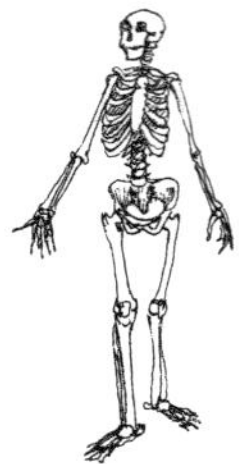

Oxidative stress and ageing

There are two main theories of ageing. One emphasizes genetic factors: primarily that our chromosomes fray at the ends and become more susceptible to damage with every cell division. The other stresses the importance of cumulative environmental damage to our bodies, mainly by those pesky free radicals.

Ageing is closely linked to oxidative stress. A lifetime of battling with oxygen, benefiting from its life-giving properties while trying to protect ourselves from its dangerous free radicals, takes its toll. Incremental damage to the molecules that make up our cells leads to mutations in DNA, disabled

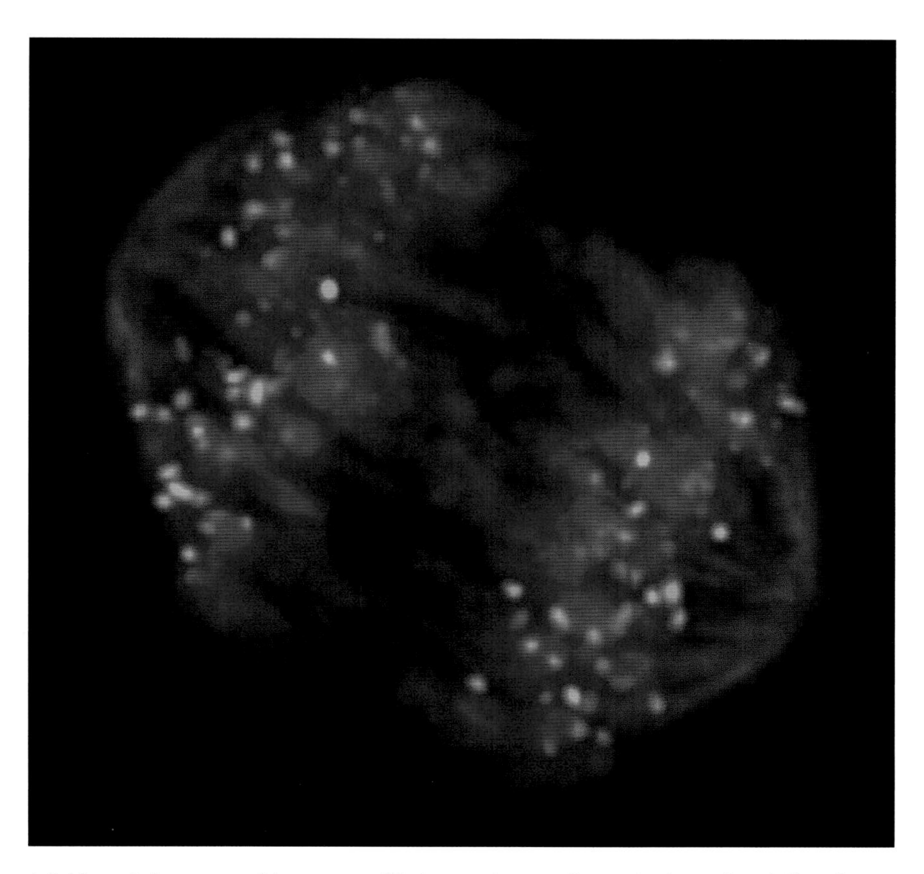

A dividing cell: the two sets of chromosomes (blue) are moving away from each other on the spindle (red), ready for the cell to split in two. Each time a cell divides, the chromosomes become more liable to damage at the ends, and errors may be introduced as the DNA is copied, producing new mutations.

enzymes, stiffened cell membranes and ultimately, the death of the cell. Around half of the damage occurring to the skin in photo-ageing is thought to be due to free radicals, and the other half, to direct damage caused by ultraviolet radiation.

Diet – specifically calorie intake – may play a strange role in oxidative stress and life expectancy. Starvation appears to increase life expectancy in rats, perhaps by increasing an animal's resistance to oxidative stress. While this is an interesting finding, be wary of what Phil Hammond calls the 'ratamorphic fallacy' – assuming that what's good (or bad) for rodents will be equally good (or bad) for humans.

However, starvation may have helped to shape how the human body functions. Food shortages are likely to have been important pressures on human populations in the past. Hunter-gatherers would probably have been quite used to going without food for one or two days, but complete famine would have been unusual, because of their variable diet and because they could move on when local food supplies ran out. With the establishment of agriculture, humans became able to manage their access to food; populations grew and cities rose. But although this looks like good risk management compared with the hunter-gatherer lifestyle, it is less flexible and very much depends on stable weather patterns, politics and peace. Storms, tyrants and war have the potential to destroy crops and stores and to plunge a country into famine. So the potential to survive famine and starvation may have then become an evolutionary advantage.

The human body responds to starvation in several ways: the metabolic rate drops to conserve energy (this also happens in dieting); non-essential physiological processes, including those related to reproduction, are switched off and physical activity declines. Starvation could potentially extend life, if slower metabolism means that cell damage is also slowed (but, conversely, it could mean that slowed processes of DNA and protein repair result in more damage and therefore a shorter lifespan). I wouldn't have thought it's worth starving yourself to find out. And, while extending life might be a desirable thing (to some extent), your quality of life isn't exactly going to be stunning if you're constantly starving yourself.

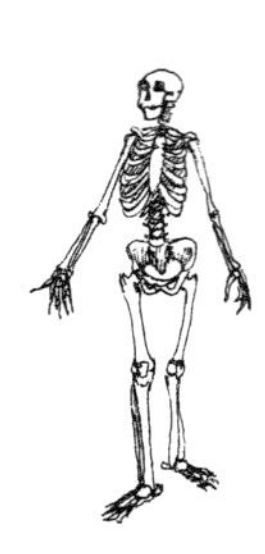

The metabolic syndrome

The 'metabolic syndrome' or 'Syndrome X' is the fastest growing medical problem in the world. Over 50 million Americans and about a third of adults in the UK have it. A syndrome is a disease which is really several diseases or conditions happening together: as our organs are interdependent, it's not surprising that problems in one have knock-on effects in the others.

A healthy, balanced diet and plenty of exercise provide the foundations of health for any organ. The metabolic syndrome illustrates what happens to your organs and your body when you undermine this foundation. A diet high in fat, especially in the wrong type of fat, leads to obesity and blood fat disorders. High levels of triglycerides, too much 'bad' (LDL) cholesterol and too little 'good' (HDL) cholesterol in the blood, means that atheroma starts to build up in the arteries. Not only that, but the blood has an increased tendency to clot. Narrowed arteries combined with a higher chance of clotting increases the risk of major damage both to the heart (through blocked coronary arteries) and the brain (through strokes). The hardened arteries lead to high blood pressure, with consequences for all organs. People with the metabolic syndrome also have insulin resistance and high blood glucose levels.

Some experts question the existence of the 'metabolic syndrome' as a thing in its own right; arguing that it's simply an extension of a familiar problem, diabetes. Others see it as a product of genes and environment, occupying that part of the spectrum where the environmental factors of poor diet, a sedentary lifestyle and obesity overcome even the best efforts of the most robust genetic constitution. For people who have some of the signs of the metabolic syndrome, but haven't developed full-blown diabetes, weight loss and exercise are more effective than drug treatment – which emphasizes the huge importance of lifestyle factors in health and disease. Whilst, undoubtedly, some people may be genetically more susceptible than others, you'll be more at risk of falling prey to the metabolic syndrome if you're overweight, physically inactive, and eat a bad diet with lots of saturated fats, trans fats, cholesterol and sugar.

Exercise, energy and the environment

The NHS Direct website presents some sobering facts about the country's health. We all know we should be doing regular exercise but it seems that only 20 per cent of us are. The most common excuse is lack of time, but I think this is misleading and unhelpful. Yes, we lead busy lives, but to imagine that we're much busier today than at any time in human history is surely apocryphal. However, something has definitely changed: as a nation, we're getting fatter and less fit.

I don't believe this is due to an innate tendency towards laziness. I think it's more to do with a struggle to adapt to a new lifestyle: specifically, to learn how to meet the challenges of living in a post-industrial society. It's not just about how busy we are – but about how we are busy. Fifty years ago, many people in the UK had jobs that involved a significant level of physical activity. Few owned cars. Today, many of us work in sedentary jobs, often sitting at a desk for much of the day. Cars are everywhere. After a long day at work and a frustrating drive home, the idea of getting in a bit of regular exercise might seem like more hard work. I don't think this is lazy, I think it's utterly understandable. But what an odd situation to find ourselves in. For thousands of years, physical activity was integral to our lives: the mainstay of how we got around and made a living. It was unavoidable and our bodies thrived on it. Now we've invented so many labour-saving devices that physical activity is no longer necessary to either getting around or making a living.

In the west, we've persuaded ourselves (or someone has persuaded us) that having a desk job is a far more priveleged way of life than using our hands to make something. Exercise has become a hobby, and has to compete with the other ways that we spend our leisure time. If only 20 per cent of us are getting enough exercise, it's not winning.

It is an interesting paradox that technological progress seems to have a negative impact on health. Does it have to be this way? I'm not a complete Luddite; there are aspects of technology, particularly in medicine, that mean we have the potential to be healthier now than ever. We understand how

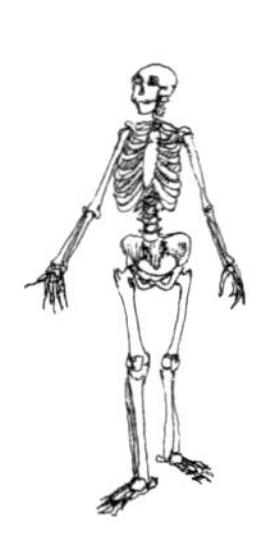

many diseases work and have come up with ingenious ways of defending ourselves against them. But it's ironic that the many benefits of technology have also made possible the unhealthy lifestyles that many of us lead. I don't think this means we should throw out all the technology and amazing medical advances and go back to living as hunter-gatherers. One thing that marks us out as humans, and underpins our success as a global species, is our adaptability to new environments. What faces us in the developed world is exactly that: a new environment. It might be one of our own making but that doesn't lessen its impact. Reducing the amount of physical activity that the body does, when it's expecting to do a lot more, is a huge change. That expectation of physical activity is deeply ingrained and is part of the way that our body and its components have evolved, so we're not going to be able to invent a pill that will somehow make up for exercising less. And would we really want to create one anyway?

We really need to tackle the problem by reintroducing exercise into our daily routines; making it something that we can't easily decide not to do if we're feeling a little under the weather or the sofa looks enticing. If you live less than a couple of miles from work – walk. If you're less than five miles away, consider cycling. Simple changes like these could have a huge impact on your health and you'll be more environmentally friendly as well. Think of it as an extension of the Gaia hypothesis: looking after your own health means you're also looking after the health of the planet. Or to put it the other way around, if you look after the earth, the earth will reward you with better health. Airborne pollution also fits in with this hypothesis: reducing the amount of pollution we pump out into the atmosphere brings health rewards for us and for the rest of the living planet.

These two Victorian ladies have incorporated exercise into their daily routines – in spite of their unsuitable clothing – and they seem to be racing ahead across Battersea Park.

The cost of a healthy life

Compared with the pre-agricultural, hunter-gatherer lifestyle that the human body evolved to cope with, it's not just levels of physical activity that might be radically different in modern western society. The hunter-gatherer diet would have been low in sodium, fat, cholesterol and refined carbohydrates and high in potassium and fibre; composed of lean meat and fish, and rich in fruits, nuts, berries and vegetables. Contrast this with the modern western diet: full of highly processed, refrigerated and fast food; it could almost have been designed to produce an epidemic of high blood pressure, coronary heart disease, strokes, renal failure, diabetes, metabolic syndrome and obesity.

There seems to be a strange relationship between the cost of food and how good it is for you. Processed foods, which take time and energy to manufacture, seem to be less expensive, and more readily available, than fresh fruit, vegetables, meat and fish. It's an equation that's hard to balance. And it produces a very disturbing social problem: fresh foods form the foundation of a healthy diet but are apparently more expensive and less accessible than less healthy, severely mucked-around and tampered-with foods. While some of us can afford to spend that bit extra on fresh, healthy stuff, some of us can't. If you're balancing a tight budget, the savings you can bank by buying processed foods, rather than the healthier raw ingredients – fish fingers rather than a bit of actual fish, burgers rather than a lean piece of meat – can be significant.

People on a lower income are less likely to eat as healthily. A study in California showed that it was up to 40 per cent cheaper to buy a 'standard basket' of foods than a 'healthy basket'. The study identified problems in supermarkets failing to stock wholewheat bread as a store brand, or to sell large packages of whole grains; low-cost store brands and bulk packages are important if you're trying to reduce the cost of groceries. While there is a general trend towards labelling healthier foods, to draw consumers' attention to them, it seems that price is much more important than healthiness. Should we be putting pressure on food retailers?

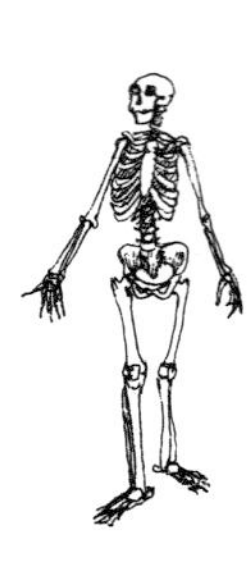

Farmers' markets are a fantastic new phenomenon: producers selling directly to consumers must surely mean that the price of fresh, local produce should come down. It may still be more expensive to buy some fresh fruit, vegetables, meat and fish direct from the producers at the moment, but we can but hope that this will change. There are also environmental reasons to support markets selling local produce: every extra mile your food travels means more energy is used for its transport and there will be a greater amount of preservatives in it.

Between issues that affect us as individuals and global issues, there's another huge problem: how do we look after other, less-fortunate, people in our society? Making sure that good food is cheap and available is a start. But having a National Health Service has to be the foundation for the health of our society; we must keep Aneurin Bevan's vision alive. Our health service may be over-burdened and under-funded, but it's the only ethical solution to health care in a modern society. As soon as it starts to get divided up, semi-privatized, out-sourced, then we run the risk of companies taking money away from it to make money for themselves, and we become consumers rather than patients.

In a recent article, a list of the medical and psycho-social features associated with asthma deaths made for sober reading: 'misuse of alcohol or drugs, psychiatric illness, denial, non-concordance with prescribed drugs, learning difficulties, income and employment difficulties, social isolation'. These are people that our modern, affluent society isn't looking after. It's not about getting expensive treatment – it's about having a role to play in society and society offering its support back to you.

Five thousand years ago, in small hunter-gatherer communities, the social issues in that list would probably have been non-existent. If what has happened over the last few millennia deserves to be called progress, this is surely how we should judge it: by how well the poorest people in our society are supported. We can still be optimistic: we can look on the present situation as part of an experiment in living with technology, and living in large conurbations, and realistically expect to progress to a condition where the outcome is better for absolutely everyone.

A healthy attitude to your organs

It's a lot easier and a lot simpler to look after your organs and the general health of your body than most news headlines, health-scare stories, food labels and 'health gurus' suggest. If you understand how your body works and what's good for it, you can approach any advice about your health with a healthy scepticism. Don't just accept the dogma; ask yourself if there's evidence that stuff works. It's important to have a healthy approach to living and eating but equally important not to feel constrained by rules and guilty if you infringe them. An obsessive approach to health is unhealthy. Watching what you eat too closely can turn into a morbid preoccupation with weight and body shape. The French paradox extends beyond the effects of wine on the heart: French people are the least worried nation in the world about their health and associate food with pleasure. In the United States, people are more food-health orientated and less food-pleasure orientated. But which is the healthier country?

Many misfortunes can affect our bodies over the course of our lives. Some might be genetically pre-determined, others might be the result of factors we have no control over. But this tour of organs has revealed just how much control we do have and that the way to give our organs the best chance of being healthy is simple: it's about what we put into our bodies and how we use them – eating a healthy diet and getting plenty of exercise is the foundation of organ and whole-body health. Remember the statistics: eight out of ten cases of heart disease, nine out of ten cases of acquired diabetes and three in ten cases of cancer could be avoided by diet and lifestyle changes.

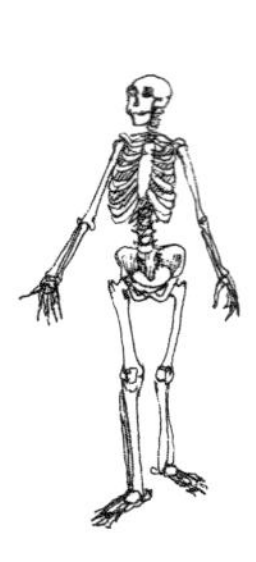

Five ways to keep your whole body healthy:

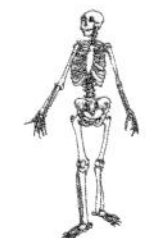

Eat a well balanced, 'Mediterranean' diet – eat things you like, enjoy eating them and remember that a little alcohol does you good.

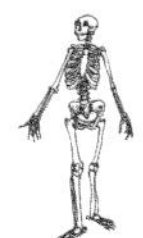

Get lots of exercise – doing things you enjoy and helping to save the planet at the same time.

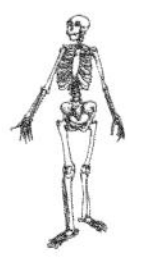

Keep your weight down – this should happen naturally if you eat well and get plenty of exercise.

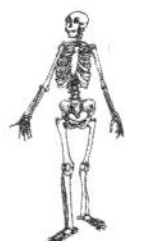

Try not to get stressed – especially about being healthy; none of this should inspire guilt.

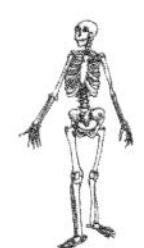

Don't smoke – bad, bad, bad: just don't do it.

AFTERWORD:

LOTIONS, POTIONS AND PILLS

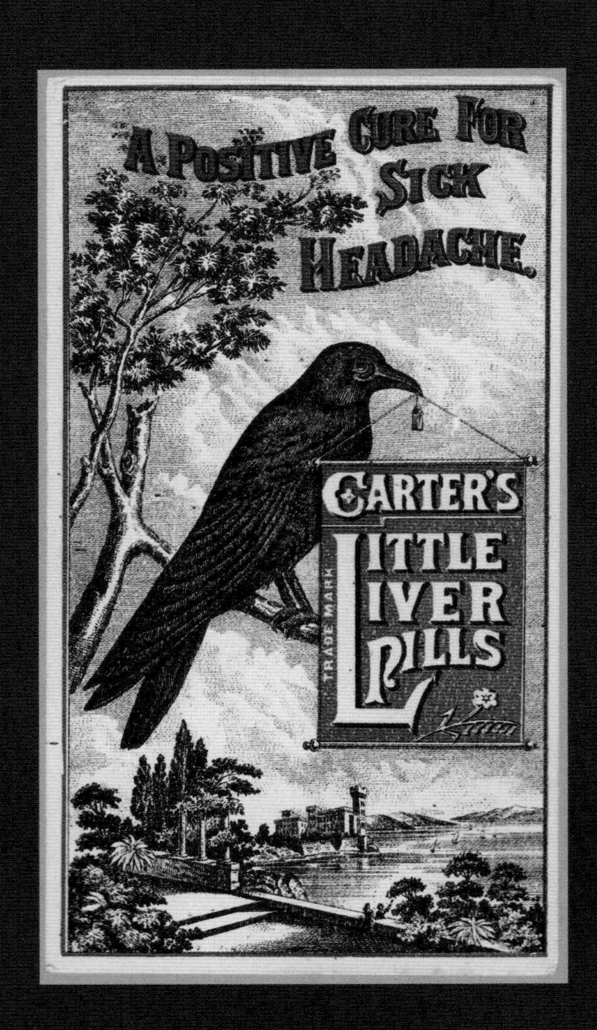

Even if you live the healthiest life on the planet, your body sometimes needs a bit of help. If you're struck down by an ailment, its severity will determine the route you take to treat it. Although some things can only be cured by letting a body-mechanic (commonly known as a surgeon) open up your body and sort it out, others need less invasive help. You might take a cup of peppermint tea to calm indigestion, go to a herbalist or seek help from your doctor if you've got a stronger or more lasting pain and need to be given medicine to ease your suffering.

Some lotions, potions and pills – over-the-counter remedies – are available in any pharmacy. You can choose between alternative medicines, including herbal remedies, and pharmaceutical drugs, often sitting on the same shelf. All of them will have been deemed safe enough to sell without a doctor getting involved. (But this isn't to say that people can't be seriously harmed by them.) They'll tell you what they do on the packet, but how do you know the claims are true?

Some lotions, potions and pills may be prescribed, whereas others are available over the counter.
Be sceptical about pills claiming to be miraculous cures or panaceas (or any advertised by crows).

Over-the-counter remedies: do they work?

Any therapy that hasn't been embraced by so-called conventional medicine is lumped together in a very unhelpful category: Complementary and Alternative Medicine (CAM). There's a problem with research into the efficacy of CAM: these remedies are generally cheap and readily available, so proving their usefulness isn't worth the investment. Some CAM practitioners further confuse the question by suggesting that their remedies are not amenable to testing with modern western science. But if you're going to take a pill, smear on a cream or have needles stuck in you, you'll want to ask: 'Does it work?' You should ask that question of any treatment, whether it's packaged as alternative or conventional medicine. Be deeply suspicious of anyone who says 'it can't be tested'; even if they believe it, they're nonetheless admitting to peddling a remedy without knowing if it works. There's been a word for exactly this type of thing since the middle of the

Visit to the quack doctor: a young viscount seeks out alternative treatment for his venereal disease in Hogarth's *Marriage à la Mode*.

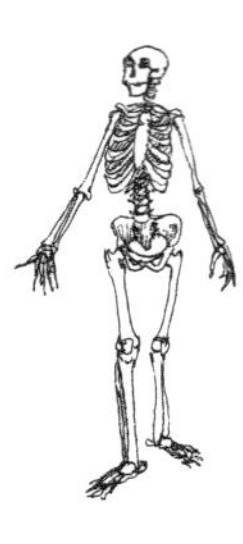

seventeenth century. It comes from a Dutch word meaning to prattle about an ointment: quacksalver.

Anyone can easily buy over-the-counter treatments without seeing a doctor. So it's important to remember that some herbal remedies interfere with medicines that you have been prescribed: for example, ephedra can reduce the effectiveness of antihypertensive drugs, *Ginkgo biloba* can increase the effect of blood-thinners like warfarin, causing bleeding, and St John's Wort makes oral contraceptives less effective. Ginseng has few side effects but can cause dangerous over-thinning of the blood if you're also taking warfarin. If you're taking any medicine, check with your doctor before taking herbal or any over-the-counter remedies.

Herbal or not, it's important to weigh the benefits of any treatment against the risks. Natural compounds, such as those in herbal remedies, aren't necessarily inherently good for you; as with any drug or remedy, you should make sure you are aware of any potential side effects. Herbal remedies are marketed as food supplements and are therefore outside the law as far as regulation of quality and safety are concerned. In 1999, a paper in *The Lancet* highlighted two cases of severe kidney disease resulting from drinking Chinese herbal tea to treat eczema.

Other remedies are equally suspect. Cough medicines seem to be no better than placebo, which is not surprising, as they are largely sugar syrup. If they do have a soothing effect, this is probably due to the sweetness rather than any active ingredient, and if so, you're likely to get just as much benefit from a home-made honey drink. For more than 60 years, people have debated the usefulness of vitamin C in preventing and treating colds. There are numerous vitamin C-containing cold cures on the market. Vitamin C might shorten the length and severity of colds but not enough to justify advising everyone to take it, and taking vitamin C after you have already developed a cold doesn't appear to have any effect at all. Several herbal remedies are used to treat or fend off the common cold, including echinacea, ginseng and astragalus; echinacea appears to have some effect, shortening the duration of a cold and lessening the symptoms, but it doesn't seem to have any preventative benefits.

The House of Lords had a really good grapple with CAM recently, with lots of experts presenting their cases on both sides. Their lordships drew up a very useful list of CAM therapies, dividing them into three groups: group 1, therapies with professional bodies to regulate them and/or generally good evidence to show that they work; group 2, therapies that often use complementary to conventional medicine, which seem to work because they induce relaxation and can be quite comforting, even if the effect is no more than that of a placebo, and group 3, therapies that present themselves as alternatives to scientific medicine, some with long traditions and some newer inventions, which, not surprisingly, don't have much scientific evidence for their efficacy. Obviously, much more research needs to be done, but for the time being, the House of Lords' list gives you a good idea of what you may reasonably expect to work and what you might very well be wasting your hard-earned cash on.

Descriptions of CAM disciplines

(from the House of Lords Science and Technology Sixth Report)

Group 1: Professionally Organized Alternative Therapies

Acupuncture (for pain relief), Chiropractic, Herbal Medicine, Homeopathy, Osteopathy

Group 2: Complementary Therapies

Alexander Technique, Aromatherapy, Bach and other Flower Remedies, Massage, Counselling Stress Therapy, Hypnotherapy, Meditation, Reflexology, Shiatsu Healing, Maharishi Ayurvedic Medicine, Nutritional Medicine

Group 3: Alternative Disciplines

With long-established and traditional systems of healthcare:

Ayurvedic Medicine, Chinese Herbal Medicine, Eastern Medicine, Naturopathy, Traditional Chinese Medicine

Other alternative disciplines:

Crystal Therapy, Dowsing, Iridology, Kinesiology, Radionics

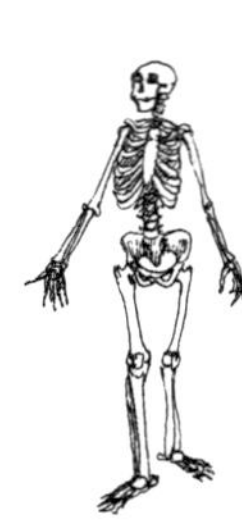

Vitamin supplements

One in four people regularly takes vitamin supplements but there's no evidence to show they do any good. If you're not getting enough of a particular vitamin, you will certainly notice the effects as you start to suffer from deficiency diseases. But in developed countries, we have access to a wide enough variety of food to get all the vitamins we need in our diet – and taking more vitamins on top won't make you any healthier.

There are two exceptions: a possible need for extra vitamin D in old age, and definite benefits from taking folic acid in pregnancy. We should get all the vitamin D we need from fruit and vegetables but there may be an argument for older people to take vitamin D supplements, along with calcium, to prevent bone loss. Self-made vitamin D is better than supplements, though we need to balance the positive effect of sunlight exposure, to make vitamin D, against the risk of skin cancer. The only vitamin supplement for which there is an unquestionably positive effect is folic acid in pregnancy, because it reduces the risk of the embryo developing neural tube defects such as spina bifida. But for most of us, most of the time, there should be no need to take vitamin supplements.

Prescribed medicines

You might think that you can place complete trust in prescribed medicines, knowing that they have passed stringent safety tests, that your doctor knows all the other medicines you're taking and will have made sure that they don't interact badly with each other, that there's good evidence that the medicine you're prescribed will effectively combat your illness and that your doctor has no vested interest in the drug.

The first two assumptions are pretty reasonable. Drugs prescribed in this country have jumped through the hoops held out by the Medicines and Healthcare Products Regulatory Agency, showing that they are safe and work for the disease they're designed to treat. Doctors are legally and ethically required to make sure you're aware of any side effects a drug might have, and to try not to prescribe cocktails of drugs that will interact badly.

But what about efficacy? Most medicines are prescribed for diseases that they've been proven to work on – but about a fifth are prescribed for diseases they're not approved for. This is called 'off-label' prescription. It sounds more dangerous than it is, because your doctor should have an understanding of how the body works and how different drugs work and be able to prescribe a drug accordingly. But it does mean there might not be any studies backing up this particular use of the drug. Your doctor should always tell you why they're prescribing a particular drug and you obviously have the final choice of whether or not to take it.

Whatever the lotion, potion or pill, over-the-counter or prescribed, you can check the evidence for whether the treatment works for yourself, on the BMJ's Best Treatments website (see page 261).

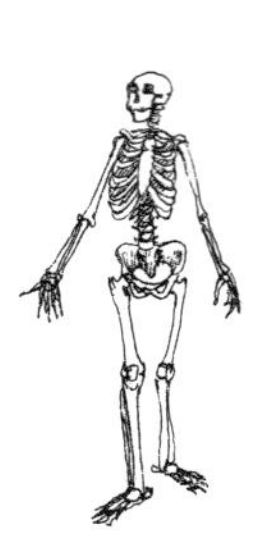

FURTHER INFORMATION

There is a great mass of information on health on the World Wide Web (or webbernet as we call it in Bristol). There's a lot of rubbish as well, so check that you are looking at information or advice coming from a reputable source (a peer-reviewed journal, a university, government or NHS website) before you decide to read/believe/follow it.

Here are some suggestions:
→For information on specific illnesses, including prevention and treatment, try the NHS Direct Online Health Encyclopaedia: www.nhsdirect.nhs.uk/index.aspx
→For a quick check on how healthy you really are – and which aspects of your diet and life you need to pay some attention to, try Harvard's Your Disease Risk: www.yourdiseaserisk.harvard.edu
→And while you're at it, you can check your life expectancy at: www.livingto100.com

For information on evidence for treatment and prevention of diseases (aimed at doctors but useful for the discerning patient):
→Bandolier: www.jr2.ox.ac.uk/bandolier
→BMJ Clinical Evidence: www.clinicalevidence.org

Other specific websites that I have found useful include:
→BMJ's Best Treatments website: www.besttreatments.co.uk/btuk/home.jsp
→Global initiative on asthma: www.ginasthma.com
→The Anaphylaxis Campaign: www.anaphylaxis.org.uk
→British Heart Foundation: www.bhf.org.uk
→British Menopause Society: www.the-bms.org
→Women's Health: www.womenshealthlondon.org.uk
→Women's Health Concern: www.womens-health-concern.org
→Breast Cancer Care: www.breastcancercare.org.uk
→BUPA: www.bupa.co.uk/health_information
→To help you keep track of medical scams and outright quackery, take a look at: www.badscience.net and Ben Goldacre's weekly column in the Saturday *Guardian:* www.quackwatch.org

→For an insider's view of the medical profession and healthy scepticism, you could do worse than to read: Phil Hammond and Michael Mosley's *Trust Me, I'm a Doctor,* published by Metro Books.
→And for more on the human machine, read Jonathan Miller's *The Body in Question*, published by Pimlico.

About the University of Bristol:
→A government-funded Centre for Excellence in Teaching and Learning was established at the University of Bristol in April 2005, building on the nationally recognized excellence in teaching within the Departments of Anatomy and Physiology. For more information, visit: http://www.bris.ac.uk/cetl/aims

Selected references

One: The Lungs

Bren F. 2002. 'Oxygen Bars: Is a breath of fresh air worth it?' *US Food and Drug Administration Consumer Magazine.* November/December 2002.

Currie GP, Devereux GS, Lee DKC, Ayres JG. 2005. 'Recent developments in asthma management.' *British Medical Journal.* 330: 585–589.

Frew AJ, Holgate ST. 2005. 'Respiratory disease.' *Clinical Medicine.* Kumar P, Clark M (eds). Elsevier Saunders.

Holgate ST. 2000. 'Science, medicine and the future: allergic disorders.' *The Lancet.* 320: 231–234.

Illi S, von Mutius E, Lau S et al. 2001. 'Early childhood infectious diseases and the development of asthma up to school age: a birth cohort study.' *British Medical Journal.* 322: 390–395.

Johnston SL, Openshaw PJM. 2001. 'The protective effect of childhood infections.' *British Medical Journal.* 322: 376–377.

Kalliomaki M, Salminen S, Arvilommi H et al. 2001. 'Probiotics in primary prevention of atopic disease: a randomised placebo-controlled trial.' *The Lancet.* 7,357 (9262): 1076–1079.

Lucas SR, Thomas AE. 2005. 'Physical activity and exercise in asthma: relevance to etiology and treatment.' *Journal of Allergy and Clinical Immunology.* 115 (5): 928–934.

Morgan WJ, Crain EF, Gruchalla RS et al. 2004. 'Results of a home-based environmental intervention among urban children with asthma.' *New England Journal of Medicine.* 351 (11): 1068–1080.

Peat JK, Mihrshahi S, Kemp AS et al. 2004. 'Three-year outcomes of dietary fatty acid modification and house dust mite reduction in the Childhood Asthma Prevention Study.' *Journal of Allergy and Clinical Immunology.* 114 (4): 807–813

Phillips M, Cataneo RN, Greenberg J et al. 2003. 'Effect of oxygen on breath markers of oxidative stress.' *European Respiratory Journal.* 21: 48–51

Platts-Mills TAE. 2005. 'Asthma severity and prevalence: an ongoing interaction between exposure, hygiene and lifestyle.' *Public Library of Science,* Medicine, www.plosmedicine.org 2(2) e34

Rees J. 2006. 'Asthma control in adults.' *British Medical Journal.* 332: 767–771.

Shore SA, Fredberg JJ. 2005. 'Obesity, smooth muscle and airway hyperresponsiveness.' *Journal of Allergy and Clinical Immunology.* 115 (5): 925–927.

Two: The Heart

Ambrose JA, Barua RS. 2004. 'The pathophysiology of cigarette smoking and cardiovascular disease: an update.' *Journal of the American College of Cardiology.* 43 (10): 1731–1717.

Anum EA, Adera T. 2004. 'Hypercholesterolaemia and coronary heart disease in the elderly: a meta-analysis.' *Annals of Epidemiology.* 14 (9): 705–721.

Burger M, Bronstrup A, Pietrzik K. 2004. 'Derivation of tolerable upper alcohol intake levels in Germany: a systematic review of risks and benefits of moderate alcohol consumption.' *Preventive Medicine.* 39: 111–127.

Camm AJ, Bunce NH. 2005. 'Cardiovascular disease.' *Clinical Medicine*. Kumar P, Clark M (eds). Elsevier Saunders.

Hannuksela ML, Ramet ME, Nissinen AET, Liisanantti MK, Savolainen MJ. 2003. 'Effects of ethanol on lipids and atherosclerosis.' *Pathophysiology*. 10: 93–103.

Hooper et al. 2001. 'Dietary fat intake and prevention of cardiovascular disease: systematic review.' *British Medical Journal*. 322: 757–763.

Houston MC. 2005. 'Nutraceuticals, vitamins, antioxidants and minerals in the prevention and treatment of hypertension.' *Progress in Cardiovascular Diseases*. 47 (6): 396–449.

Iestra JA, Kromhout D, van der Schouw YT, Grobbee DE, Boshuizen HC, van Staveren WA. 2005. 'Effect size estimates of lifestyle and dietary changes on all-cause mortality in coronary artery disease patients: a systematic review.' *Circulation*. 112 (6): 924–934.

Ferrieres J. 2004. 'The French Paradox: lessons for other countries.' *Heart*. 90: 107–111.

Kromhout D, Menotti A, Kesteloot H, Sans S. 2002. 'Prevention of Coronary Heart Disease by Diet and Lifestyle.' *Circulation*. 105: 893–898.

McGavock JM, Anderson TJ, Lewanczuk RZ. 2006. 'Sedentary lifestyle and antecedents of cardiovascular disease in young adults.' *American Journal of Hypertension*. 19 (7): 701–707.

Obarzanek et al. 2003. 'Individual blood pressure responses to changes in salt intake.' *Hypertension*. 42: 459–467

Oomen CM et al. 2001. 'Association between trans-fatty acid intake and 10-year risk of coronary heart disease in the Zutphen Elderly Study: a prospective population-based study.' *The Lancet*. 357: 746–751.

Opie LH, Commerford PJ, Gersh BJ. 2006. 'Controversies in stable coronary artery disease.' *The Lancet*. 367: 69–78.

Padwal R, Straus SE, McAlister FA. 'Evidence-based management of hypertension: cardiovascular risk factors and their effects on the decision to treat hypertension: evidence based review.' *British Medical Journal*. 322: 977–980.

Three: The Stomach and Intestines

Advertising Standards Organisation. 'Adjudication: Allergy Testing Service.' May 1999, number 96.

Clark ML, Silk DBA. 2005. 'Gastrointestinal disease.' *Clinical Medicine*. Kumar P, Clark M (eds). Elsevier Saunders.

Ernst E. 1997. 'Colonic irrigation and the theory of autointoxication: a triumph of ignorance over science.' *Journal of Clinical Gastroenterology*. 24 (4): 196–198.

Frezza EE, Wachtel MS, Chiriva-Internati M. 2006. 'Influence of obesity on the risk of developing colon cancer.' *Gut*. 55 (2): 285–291.

Guarner F, Malagelada J-R. 2003. 'Gut flora in health and disease.' *The Lancet*. 361: 512–519.

Jacobsen MB, Aukrust P, Kittang E et al. 2000. 'Relation between food provocation and systemic immune activation in patients with food intolerance.' *The Lancet*. 356: 400–401.

Kailasapathy K, Chin J. 2000. 'Survival and therapeutic potential of probiotic organisms with reference to *Lactobacillus acidophilus* and *Bifidobacterium* spp.' *Immunology and Cell Biology*. 78:80–88.

Loftus EV. 2004. 'Clinical epidemiology of inflammatory bowel disease: incidence, prevalence and environmental influences.' *Gastroenterology*. 126: 1504–1517.

Quigley EMM, Quera R. 2006. 'Small intestinal bacterial overgrowth: roles of antibiotics, prebiotics and probiotics.' *Gastroenterology*. 130: S70–S90.

Sazawal S, Hiremath G, Dhingra U, Malik P, Deb S, Black RE. 2006. 'Efficacy of probiotics in prevention of acute diarrhoea: a meta-analysis of masked, randomised, placebo-controlled trials.' *Lancet Infectious Diseases*. 6: 374–382.

Sicherer SH. 2002. 'Food allergy.' *The Lancet*. 360: 701–710.

Swagerty DL, Walling AD, Klein RM. 2003. 'Lactose intolerance.' *American Family Physician*. 65 (9): 1845–1850.

Weitz J, Koch M, Debus J, Hohler T, Galle PR, Buchler MW. 2005. 'Colorectal cancer.' *The Lancet*. 365 (9464): 153–165.

Yamaguchi N, Kakizoe T. 2001. 'Synergistic interaction between *Helicobacter pylori* gastritis and diet in gastric cancer.' *The Lancet Oncology*. 2: February 2001.

Four: The Liver

Burger M, Bronstrup A, Pietrzik K. 2004. 'Derivation of tolerable upper alcohol intake levels in Germany: a systematic review of risks and benefits of moderate alcohol consumption.' *Preventive Medicine*. 39: 111–127.

Burroughs AK, Westaby D. 2005. 'Liver, biliary tract and pancreatic disease.' *Clinical Medicine*. Kumar P, Clark M (eds). Elsevier Saunders.

Chuang CZ, Martin LF, LeGardeur BY et al. 2001. 'Physical activity, biliary lipids and gallstones in obese subjects.' *American Journal of Gastroenterology*. 96 (6): 1860–1865.

Leitzman MF, Stampfer MJ, Willett WC et al. 2002. 'Coffee intake is associated with lower risk of symptomatic gallstone disease in women.' *Gastroenterology*. 123 (6): 1823–1830

Pittler MH, Verster JC, Ernst E. 2005. 'Interventions for preventing or treating alcohol hangover: systematic review of randomised controlled trials.' *British Medical Journal*. 331: 1515–1518.

Portincasa P, Moschetta A, Palasciano G. 2006. 'Cholesterol gallstone disease.' *The Lancet*. 368: 230–239.

Room R. 'British livers and British alcohol policy.' *The Lancet*. 367: 10–11.

Five: The Pancreas

Daly ME. 2004. 'Extending the use of the glycaemic index: beyond diabetes?' *The Lancet*. 364: 736–737.

Eriksson J, Lindstrom J, Tuomilehto J. 2001. 'Potential for the prevention of type 2 diabetes.' *British Medical Bulletin*. 60: 188–199.

Gale EAM, Anderson JV. 2005. 'Diabetes mellitus and other disorders of metabolism.' *Clinical Medicine*. Kumar P, Clark M (eds). Elsevier Saunders.

Nicholls D, Viner R. 2005. 'Eating disorders and weight problems.' *British Medical Journal*. 330: 950–953.

WHO 2003. 'Diet, nutrition and the prevention of chronic diseases.' WHO technical report series: 916.

Williams DE et al. 1999. 'Frequent salad vegetable consumption is associated with a reduction in the risk of diabetes mellitus.' *Journal of Clinical Epidemiology*. 52: 329–335.

Six: The Kidneys and Bladder

Berghmans LC, Hendriks HJ, Bo K et al. 1998. 'Conservative treatment of stress urinary incontinence in women: a systematic review of randomised clinical trials.' *British Journal of Urology*. 82: 181–191.

Bruyere F. 2006. 'Use of cranberry in chronic urinary tract infections.' *Medicine et Maladies Infectieuses*. 36 (7): 358–363.

Foo LY, Lu Y, Howell AB et al. 2000. 'A-Type proanthocyanidin trimers from cranberry that inhibit adherence of uropathogenic P-fimbriated *Escherichia coli*.' *Journal of Natural Products*. 63 (9): 1225–1228.

Kunin CM, Evans C, Bartholomew D, Bates DG. 2002. 'The antimicrobial defense mechanisms of the female urethra: a reassessment.' *Journal of Urology*. 168 (2): 413–419.

Moe OW. 2006. 'Kidney stones: pathophysiology and medical management.' *The Lancet*. 367: 333–344.

Yaqoob M. 2005. 'Renal disease.' *Clinical Medicine*. Kumar P, Clark M (eds). Elsevier Saunders.

Seven: The Reproductive Organs

Avorn J, Monane M, Gurwitz JH et al. 1994. 'Reduction of bacteriuria and pyuria after ingestion of cranberry juice.' *Journal of the American Medical Association*. 271: 751–754.

Doherty L, Fenton KA, Jones J et al. 2002. 'Syphilis: old problem, new strategy.' *British Medical Journal*. 325: 153–156.

Finch RG, Moss P, Jeffries DJ, Anderson J. 2005. 'Infectious diseases, tropical medicine and sexually transmitted diseases.' *Clinical Medicine*. Kumar P, Clark M (eds). Elsevier Saunders.

Gilson RJC, Mindel A. 2001. 'Recent advances: sexually transmitted infections.' *British Medical Journal*. 322: 1160–1164.

Hickey M, Davis SR, Sturdee DW. 2005. 'Treatment of menopausal symptoms: what shall we do now?' *The Lancet*. 366: 409–421.

Hilton E, Isenberg HD, Alperstein P et al. 1992. 'Ingestion of yoghurt containing *Lactobacillus acidophilus* as a prophylaxis for candidal vaginitis.' *Annals of Internal Medicine*. 116 (5): 353–357.

Latthe P, Latthe M, Say L et al. 2006. 'WHO systematic review of prevalence of chronic pelvic pain: a neglected reproductive health morbidity.' *Biomed Central Public Health*. 6: 177

Lister TA, Gallagher CJ. 2005. 'Malignant disease.' *Clinical Medicine*. Kumar P, Clark M (eds). Elsevier Saunders.

McPherson K, Steel CM, Dixon JM. 2000. 'Breast cancer – epidemiology, risk factors and genetics.' *British Medical Journal*. 321: 624–628.

Moyad MA, Carroll PR. 2004. 'Lifestyle recommendations to prevent prostate cancer, part II: time to redirect our attention?' *Urologic Clinics of North America*. 31 (2): 301–311.

Pirotta M, Gunn J, Chondros P et al. 2004. 'Effect of *Lactobacillus* in preventing post-antibiotic vulvovaginal candidiasis: a randomised controlled trial.' *British Medical Journal*. 329 (7): 548.

Proctor M, Farquhar C. 2006. 'Diagnosis and management of dysmenorrhoea.' *British Medical Journal*. 332: 1134–1138.

Rieck G, Fiander BM. 2006. 'The effect of lifestyle factors on gynaecological cancer.' *Best Practice and Research Clinical Obstetrics and Gynaecology*. 20, 2: 227–251.

Reid G, Bocking A. 2003. 'The potential for probiotics to prevent bacterial vaginosis and preterm labour.' *American Journal of Obstetrics and Gynecology*. 189 (4): 1202–1208.

Rosen RC, Friedman M, Kostis JB. 2005. 'Lifestyle management of erectile dysfunction: the role of cardiovascular and concomitant risk factors.' *The American Journal of Cardiology*. 96 (suppl): 76M–79M.

Eight: The Brain

Clarke CRA. 2005. 'Neurological disease.' *Clinical Medicine*. Kumar P, Clark M (eds). Elsevier Saunders.

Clarke R et al. 1998. 'Folate, vitamin B12 and serum total homocysteine levels in confirmed Alzheimer disease.' *Archives of Neurology*. 55: 1449–1455.

Lawlor DA, Hopker SW. 2001. 'The effectiveness of exercise as an intervention in the management of depression: systematic review and meta-regression analysis of randomised controlled trials.' *British Medical Journal*. 322: 1–8.

Le Bars PL, Katz MM, Berman N et al. 1997. 'A placebo-controlled, double-blind, randomized trial of an extract of *Ginkgo biloba* for dementia.' *Journal of the American Medical Association*. 278: 1327–1332.

Scott J. 2006. 'Depression should be managed like a chronic disease.' *British Medical Journal*. 332: 985– 986.

Small GW. 2002. 'What we need to know about age-related memory loss.' *British Medical Journal*. 324: 1502–1505.

Nine: The Eye

Azzam N, Dovrat A. 2004. 'Long-term lens organ culture system to determine age-related effects of UV irradiation on the eye lens.' *Experimental Eye Research*. 79: 903–911.

Blehm C, Vishnu S, Khattak A et al. 2005. 'Computer vision syndrome: a review.' *Survey of Ophthalmology*. 50 (3): 253–262.

Chopdar A, Chakravarthy U, Dinesh V. 2003. 'Age-related macular degeneration.' *British Medical Journal*. 326: 485–488.

USDA Nutrient Data Laboratory. 1998. Carotenoid Database. www.nal.usda.gov/fnic/foodcomp/Data/car98/car98.html

Wu J, Seregard S, Algvere PV. 2006. 'Photochemical damage of the retina.' *Survey of Ophthalmology*. 51: 461–481.

Ten: The Skin

Bastuji-Garin S, Diepgen TL. 2002. 'Cutaneous malignant melanoma, sun exposure and sunscreen use: epidemiological evidence.' *British Journal of Dermatology*. 146 (61): 24–30.

Buck P, 2004. 'Skin barrier function: effect of age, race and inflammatory disease.' *International Journal of Aromatherapy*. 14: 70–76.

Johnson AW. 2004. 'Overview: fundamental skin care – protecting the barrier.' *Dermatologic Therapy*. 17: 1–5.

Harding CR 2004. 'The stratum corneum: structure and function in health and disease.' *Dermatologic Therapy*. 17: 6–14.

Hiom S. 2006. 'Public awareness regarding UV risks and vitamin D – the challenges for UK skin cancer campaigns.' *Progress in Biophysics and Molecular Biology*. 92: 161–166.

Laube S. 2004. 'Skin infections and ageing.' *Ageing Research Reviews*. 3: 69–89.

Loden M. 2004. 'Do moisturizers work?' *Journal of Cosmetic Dermatology*. 2: 141–149.

Orton DI, Wilkinson JD. 'Cosmetic allergy: incidence, diagnosis and management.' *American Journal of Clinical Dermatology*. 5 (5): 327–337.

Rabe JH, Mamelak AJ, McElgunn PJS, Morison WL, Sauder DN. 2006. 'Photoaging: mechanisms and repair.' *Journal of the American Academy of Dermatology*. 55 (1): 1–19.

Zitterman. 2006. 'Vitamin D and disease prevention.' *Progress in Biophysics and Molecular Biology*. 92 (1): 39–48.

Eleven: The Whole Body

Cummings JH, Bingham SA. 1998. 'Diet and the prevention of cancer.' *British Medical Journal*. 317: 1636–1640.

Foreyt JP. 2005. 'Need for lifestyle intervention: how to begin.' *American Journal of Cardiology*. 96, 4 (1): 11–14.

Jetter KM, Cassady DL. 2006. 'The availability and cost of healthier food alternatives.' *American Journal of Preventive Medicine*. 30 (1): 38–44.

Nicholls D, Viner R. 2005. 'Eating disorders and weight problems.' *British Medical Journal*. 330: 950–953.

Prentice AM. 2005. 'Starvation in humans: evolutionary background and contemporary implications.' *Mechanisms of Ageing and Development*. 126: 976–981.

Willcox JK, Ash SL, Catignani GL. 2004. 'Antioxidants and the prevention of chronic disease.' *Critical Reviews in Food Sciences and Nutrition*. 44: 275–295.

Afterword

Bender DA. 2002. 'Daily doses of multivitamin tablets.' *The Lancet*. 325: 173–174.

Conover EA. 2003. 'Herbal agents and over-the-counter medications in pregnancy.' *Best Practice and Research Clinical Endocrinology and Metabolism*. 17 (2): 237–251.

Ernst E. 2000. 'Herbal medicines: where is the evidence?' *British Medical Journal*. 321: 396–398.

House of Lords Select Committee on Science and Technology Sixth Report. www.publications.parliament.uk

Schroeder K, Fahey T. 2002. 'Systematic review of randomised controlled trials of over-the-counter cough medicines for acute cough in adults.' *British Medical Journal*. 324: 1–6.

Tattersall M, Kerridge I. 2006. 'The drug industry and medical professionalism.' *The Lancet*. 367: 28.

US Department of Agriculture Agricultural Research Service. 'Can antoxidant foods forestall ageing?' *Food and Nutrition Research Briefs*. April 1999. www.ars.usda.gov/is/np/fnrb/fnrb499.htm

INDEX

Page numbers in **bold** refer to the main discussions of specific organs; page numbers in *italics* refer to the illustrations.

PICTURE CREDITS

The publishers wish to thank the following picture libraries for permission to reproduce images in this book:

© Bridgeman Art Library – p.161, p.256

© Mary Evans Picture Library – p.6, p.18, p.37, p.249

© Mediscan – p.15(l), p.62

© Science Photo Library – p.12, p.14(r), p.20, p.22, p.28, p.30, p.35, p.38, p.44, p.46, p.52, p.56, p.67, p.72, p.76, p.77, p.82, p.86, p.89, p.90, p.94, p.96, p.100, p.103, p.104, p.106, p.110, p.112, p.114, p.117, p.119, p.121, p.124, p.126, p.128, p.132, p.134, p.135, p.137, p.142, p.143, p.146, p.148, p.150, p.154, p.156, p.158, p.159, p.163, p.170, p.172, p.175, p.178, p.179, p.180, p.181, p.182, p.184, p.187, p.192, p.196, p.198, p.199, p.200, p.203, p.204

© TopFoto – p.61, p.206, p.210, p.212, p.217, p.218, p.221, p.225, p.226, p.231, p.234, p.236, p.239, p.241, p.242

© Wellcome Trust Medical Photographic Library – p.13, p.16, p.32, p.70, p.80, p.176, p.194, p.245, p.254

The publishers also wish to thank the anatomy department at King's College London for the use of one of their skeletons in the photography for this book.

Acknowledgements

Enormous thanks to Hilary and Luigi. I would like to thank Richard Atkinson and the rest of the fantastic team who have worked on the book: Natalie Hunt, Juliet Davis, Emma Smith, Alex Smith, Jim Smith, David Atkinson, Penny Edwards, Kate Bland, Erica Jarnes and Phil Beresford. Also, thank you to Ann Grand for her editing expertise and to Oxford Designers and Illustrators Limited for their elegant diagrams.

Thank you to all the team at BBC Birmingham involved in making the series, and in particular, Gill Tierney and Kathy Myers, for turning an embryonic idea about an anatomy television programme into a living and breathing reality.

And finally, thank you to Dave Stevens for all his patience and love.

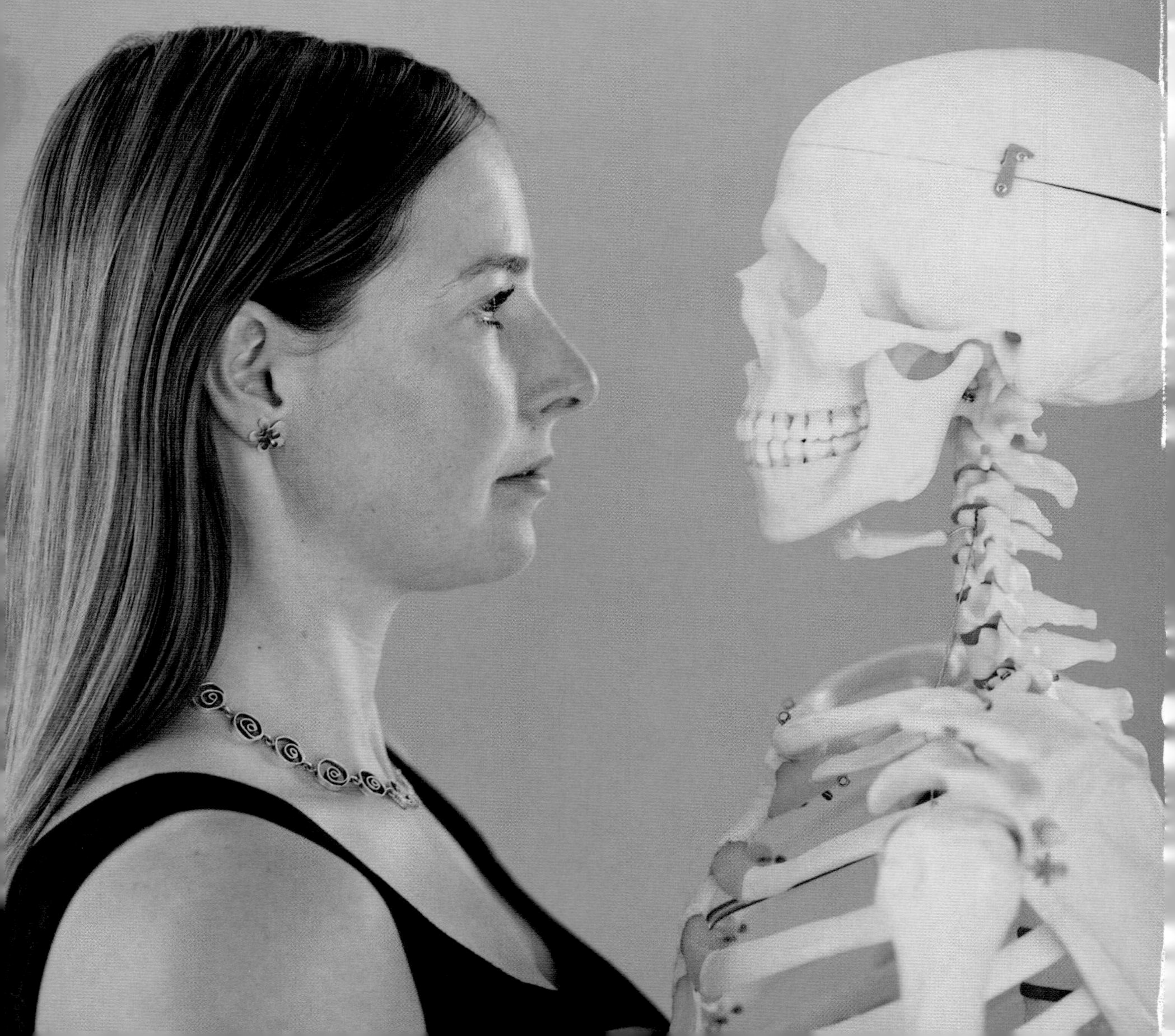